Establishing the Environmental Flow Regime for

Establishing the Environmental Flow Regime for the Middle Zambezi River

DISSERTATION

Submitted in fulfillment of the requirements of
the Board for Doctorates of Delft University of Technology
and
of the Academic Board of the UNESCO-IHE
Institute for Water Education
for
the Degree of DOCTOR
to be defended in public on
Monday, 13 June 2016, at 12:30 hours
in Delft, the Netherlands

by

Elenestina Mutekenya MWELWA
Master of Science, Rhodes University, South Africa
born in Mporokoso, Zambia

This dissertation has been approved by the
promotor: Prof.dr.ir. A.E. Mynett and
copromotor: Dr.ir. A. Crosato

Composition of the doctoral committee:

Chairman	Rector Magnificus TU Delft
Vice-Chairman	Rector UNESCO-IHE
Prof.dr.ir. A.E. Mynett	UNESCO-IHE / TU Delft, promotor
Dr.ir. A. Crosato	UNESCO-IHE / TU Delft, copromotor
Independent members:	
Prof.dr.ir. S.N. Jonkman	TU Delft
Prof.dr. K.A. Irvine	UNESCO-IHE / Wageningen University
Prof.dr. N.G. Wright	De Montfort University, UK
Dr. F. Hughes	Anglia Ruskin University, UK
Prof.dr.ir. H.H.G. Savenije	TU Delft, reserve member

Published by:
CRC Press/Balkema
PO Box 11320, 2301 EH Leiden, The Netherlands
Pub.NL@taylorandfrancis.com
www.crcpress.com – www.taylorandfrancis.com
ISBN 978-1-138-03180-7

Dedication

To the loving memory of a dear husband and friend, gone too early, the late Chato Fackson Mwelwa. Your encouragement in my professional pursuits are still valued.

Acknowledgements

As an African proverb says that "it takes a village to raise a child", in the same way, I would say that it has taken a dedicated and supportive team of mentors, family members and friends for me to have made it this far. For quite often, the road would get so crowded that I would fail to see the way ahead, but within the supportive team, would be the wise and encouraging words to keep me going. I am glad that within this thesis is a section where I can express my sincere gratitude to my supportive team members that have seen me to this end.

First and foremost, I would like to posthumously thank my late husband Chato, who believed in me and encouraged me to have courage that all would be possible and that I was capable of handling even the toughest tasks. His belief in me, build my inner confidence to be able to take on difficult and worthwhile tasks, even when others would not approve or see the essence. I am truly grateful that I was given an opportunity to live with a great man who had a strong desire to see his spouse succeed.

Around me have also been great friends, who did believe that I was capable to take on the journey and complete it. Of these friends, is Bishop Chali Kasonde, who after I had completed my MSc visited me and said now what was remaining was a PhD. To be honest, at the time he said that, I had no idea that I would ever venture into tackling this task. A number of years passed, but true to his word, here I am jumping over the hurdle, I am truly grateful for the encouragement and support.

Having had a responsibility to look after a large family (nicknamed the 144,000), quite often, it was very difficult to leave home and spend several months away in the Netherlands. During my absence, the management of the big family would then fall on unsuspecting family members, for whom I am so grateful.

The journey has been long, but along the way, God did place strategically helpers to encourage me and see me to the next level. The start was so difficult, but it only took a few minutes of chatting with Prof. Nigel Wright, and my energy and strength would be revived, I am so grateful that his guidance and support has continued to the very end. The last time we met briefly, he encouraged me that the next time we would meet would be during my defence. Dr. Lindsay Beevers, was one who participated in the selection process when the position for this PhD was advertised under the power2flow project, her support and guidance was so valuable. I was particularly glad that one day, she was able to confide in me that she was happy that the selection team picked me for the PhD position, which was quite encouraging indeed. I am so grateful to the power2flow project team for making the financial resources available for the work.

In this team of helpers, I would be failing if I do not particularly mention that the reason that I am able to complete this task is indeed Dr. Alessandra Crosato's personal effort. I do believe that she did fight battles on my behalf, to just ensure that I stayed on course, Oh what a lady! I am so grateful that God put her in this important team to just ensure that I complete the task. Her encouraging words and her personal effort to keep me going were often the reason I would wake up and just keep going, when the going was rough. There was a time when in literal sense she would hold my hand and cheer me on. I am so grateful for the professional meticulous way she would provide the guidance, she has been such a blessing to me.

In this team is also Prof. Arthur Mynett. I am so grateful that he did believe that this could be done and that when resources were a constraint, he would make personal effort to find resources to keep me going in order to cross the most difficult patches. The encouragement and guidance provided was so good and up to date, I would fail to just say he was one placed on the way to help me cross over and get the work done. Of all, the effort, when my progress was not sufficient back home, the suggestion was to camp in the Netherlands for several months to ensure I completed writing the thesis. The writing camp made such a big difference, I am so grateful indeed.

Having field work in a very challenging environment, as a lady, I always needed helpers. I would particularly express my gratitude to one, Mr. Christopher Kaniki, who proved himself as a very helpful field support. For the successful river survey work, I am so grateful for the helping hand and expertise provided by one, Mr. Sam Mwale. I also wish to thank the Gwabi Lodge Management, who were always welcoming and for always making the best boat and the best coxswain, Mr. Forbus available to take me up and down the Zambezi River.

I wish to also thank my employers ZESCO Management for providing resources for field work and for giving me an opportunity to take leave of absence and fulfill the requirements of this PhD.

In all, it has taken a great team to see me through, for all this I give God the glory and honor. With God, all things are possible indeed.

Delft, 13th June 2016

Elenestina Mutekenya Mwelwa

Summary

The Zambezi River and its tributaries constitute one of Southern Africa's most important natural resources. The Middle Zambezi is located in the central part of the Zambezi Basin, which has a total surface area of 1,357,000 km^2 and extends over portions of eight Southern African Countries. From the rich biodiversity that this sub-catchment supports, both Zambia and Zimbabwe have established National Parks, with Mana Pools National Park, Sapi and Chewore safari areas being designated as UNESCO World Heritage Site in 1984. The habitat sustenance depends on the river channels and the associated morphological features with the flood and recession interactions, whose modification can lead to negative environmental consequences.

The area is located downstream of three hydropower schemes. The Victoria Falls run-of-river power station, was commissioned in three phase, 1938, 1969 and 1972 and has an installed capacity of 108 MW. The Kariba scheme comprises the Kariba Dam, and two power stations, the Kariba South Bank, commissioned in 1958 with an installed capacity of 750 MW, and the Kariba North Bank, commissioned in 1976 with an installed capacity of 1020 MW. The Kafue Gorge power station, built on the River Kafue, a tributary of the Zambezi with an installed capacity of 990 MW, was commissioned in 1977 and is served by two reservoirs, the Itezhi-tezhi (main storage) and the Kafue Gorge. The Middle Zambezi River ends in the reservoir created by Cahora Bassa Dam with a power station of 2,075 MW, which is managed by Hidroeléctrica de Cahora Bassa (Mozambique).

The hydropower dam operators' focus is on optimizing hydropower generation, leading to lack of flooding, as most of the flows released are turbine discharges. In the last decades, the floodplain riparian tree, the *Faidherbia albida*, which is of vital importance for the local wild life, has shown a worrying decrease in its regeneration rates, so that young trees can only be found on the river islands, whereas the river banks host deteriorating old trees.

The research work in the area had the key objective to establish the environmental flow regime for the Middle Zambezi reach which minimises the impact of the upstream hydropower schemes on the river environment. This requires the assessment of the effects of regulated water flow on the short and long term morphological evolution of the river, floodplain vegetation and grazing animals, as well as their interactions. The assessment requires the study of the pre-impoundment and post-impoundment states of the Middle Zambezi and future state scenarios.

A combination of research methodologies was used in the research. This comprises numerical modelling of surface and ground water flows, morphodynamic changes, field investigations, historical analysis of discharge series, including the discharge regime before and after the construction of the two dams of Kariba and Kafue, as well as laboratory investigations on plant growth rates. The work also included both satellite and map analyses to study the historical morphological changes. Detailed field vegetation surveys, as well as hydrodynamic, ground water and sediment data collections were undertaken to understand how floodplain vegetation interacts with the hydro-morphological system. The *Faidherbia albida* tree was chosen as a biological indicator.

The results show three distinguishable hydrodynamic periods, being: pre-Kariba; post-Kariba 1, with only one power station operational, and post-Kariba 2, with three power stations operational: Kariba South Bank, Kariba North Bank and Kafue Gorge. The comparison between hydrodynamic simulations of the river flow shows that the current dam operations have completely altered the natural hydrological rhythm, in particular floodplain inundation. Moreover, the pre-Kariba dry season flows of 100-200 m^3/s are now 1,000-1,500 m^3/s. In addition, the observed river channel widening phenomenon appears linked to the sudden reduction of river water levels due to the sudden closure of dam flood gates.

The vegetation surveys indicate that in some areas of the floodplain, there are only old trees of aged 50 years or more, with a lot of die offs. The *F. Albida* seed was observed to be very robust and easily germinating in moist conditions, but seedlings seem to fail to establish. Growth of younger trees was only observed in islands inaccessible to animals most of the year. The leaves, tree bark and fruits belong to the food supply of most wild animals, so young trees are highly prone to be browsed as soon as they grow. The regeneration survival of the tree therefore lies in longer flood residence over portions of the floodplain making the areas inaccessible to animals to allow for regeneration.

To save the *F. albida* tree from being completely exterminated from this floodplain, which is obtained by keeping the animals away from some sections of the floodplain for about six months, the following two-pronged environmental flow regime is proposed: the dam operators facilitate a deliberate release of a moderate flood of 5,800 m^3/s once in 5 years, which should have duration of 5 to 6 weeks, in the months of February to March. Then, the spillway gates closing is phased over a period of 3 to 4 weeks to ensure that the flood and ground water table recession is slowed down to keep the floodplain wet enough until the months of May and June. This would allow for seed germination and the seedlings to be established. The phasing of spillway gate closure would also mitigate the current channel widening phenomenon which results from excessive bank erosion caused by the sudden closure of the dam flood gates.

Samenvatting

De Zambezi met zijn zijrivieren is een van de belangrijkste natuurgebieden van Zuidelijk Afrika. De Midden-Zambezi is centraal gelegen met een oppervlakte van $1,357,000$ km^2 dat zich uitstrekt over acht landen. Vanwege de rijke biodiversiteit in dit gebied hebben zowel Zambia als Zimbabwe hier nationale parken ingericht waarvan Mana Pools, Sapi en Chewore sinds 1984 tot het UNESCO werelderfgoed behoren. De leefgebieden van vele plant- en diersoorten aldaar zijn sterk afhankelijk van waterstandvariaties en morfologische veranderingen in de (zij)rivieren. Veranderingen daarin kunnen negatieve gevolgen hebben voor het natuurlijk milieu.

De Midden-Zambezi is stroomafwaarts gelegen van drie waterkrachtcentrales. De Victoria Falls waterkrachtcentrale is in drie fasen aangelegd in 1938, 1969 en 1972 en heeft een geïnstalleerd vermogen van 108 MW. De Kariba centrale omvat de Kariba Dam en twee krachtcentrales, de Kariba South Bank (aangelegd in 1958 met 750 MW vermogen) en de Kariba North Bank (aangelegd in 1976 met een vermogen van 1020 MW). De Kafue Gorge waterkrachtcentrale bevindt zich in de rivier de Kafue, een zijrivier van de Zambezi, met een geïnstalleerd vermogen van 990 MW, is in 1977 aangelegd en wordt gevoed door twee reservoirs, the Itezhi-tezhi en de Kafue Gorge. De Midden-Zambezi mondt uit in het stuwmeer van de Cahora Bassa Dam met een krachtcentrale van 2,075 MW dat wordt beheerd door Hidroeléctrica de Cahora Bassa in Mozambique.

Het doel van de beheerders van de waterkrachtcentrales is gericht op het optimaliseren van de elektriciteitsopwekking. Hierbij worden variaties in waterstanden afgevlakt door het stuwmeer met als gevolg zeer gelijkmatig verdeelde debieten uit de turbines en nauwelijks meer natuurlijke overstromingen. Vandaar dat gedurende de afgelopen decennia de aanwezigheid van de boom *Faidherbia albida*, van vitaal belang voor de lokale wilde dieren, aanzienlijk is achteruit gegaan en alleen nog maar wordt aangetroffen op eilandjes in de rivier en niet meer langs de oevers.

Het doel van dit promotie onderzoek was om vast te stellen welk natuurlijke variaties in het stroomgebied van de Midden-Zambezi nodig zijn om de negatieve gevolgen van de stroomopwaarts gelegen waterkrachtcentrales te beperken. Daartoe is onderzoek gedaan naar de effecten van gereguleerde stromingen op de korte en lange termijn morfologische veranderingen in de rivier, de vegetatie in de uiterwaarden, en de grazende dieren – en van de interacties daartussen. Daartoe is onderzocht hoe de Midden-Zambezi er uit zag voor en na de aanleg van de waterkrachtcentrales, en welke maatregelen kunnen worden getroffen om nadelige effecten te verzachten.

In dit onderzoek is een combinatie van methoden toegepast, waaronder het uitvoeren van historische analyses van stromingen in de rivier voor en na de constructie van de Kariba en Kafue dam; het numeriek modelleren van oppervlakte- en grondwater-stromingen en de daardoor veroorzaakte morfologische veranderingen; het uitvoeren van veldonderzoek; alsmede laboratorium proeven naar het aangroeien van planten.

Andere onderdelen betroffen het analyseren van satellietbeelden en oude kaarten om de morfologische veranderingen te bestuderen. Gedetailleerde veldstudies samen met gegevens over oppervlakte- en grondwaterstromingen en gegevens omtrent sediment-samenstelling gaven inzicht in de interactie tussen vegetatie in de uiterwaarden en het hydro-morfologisch gedrag van het systeem. De boomsoort *Faidherbia albida* is gekozen als biologische indicator.

De resultaten laten zien dat er drie onderscheidende hydrodynamische perioden zijn geweest: pre-Kariba; post-Kariba 1 (met een operationele waterkrachtcentrale); en post-Kariba 2 (met drie operationele centrales: Kariba South Bank, Kariba North Bank en Kafue Gorge). Een vergelijk tussen hydrodynamische simulaties van de rivierstromen laat zien dat het huidige beheer van de reservoirs het natuurlijke hydrologische ritme volledig verstoord hebben, met name van het overstromen van de uiterwaarden. De debieten uit het pre-Kariba droge seizoen van destijds 100-200 m^3/s bedragen thans 1,000-1,500 m^3/s. Bovendien lijkt de waargenomen oevererosie volledig te worden veroorzaakt door het plotseling verlagen van de waterstanden door het te snel sluiten van de schuiven in de dam.

Uit het vegetatieonderzoek blijkt dat in sommige delen van het stroomgebied alleen nog bomen staan van 50 jaar of ouder, waarvan een groot deel niet meer leeft. Het zaad van de *F. Albida* blijkt zeer robust en makkelijk te ontkiemen in vochtige omstandigheden, maar de zaailingen willen zich maar moeizaam vestigen. Jongere bomen zijn alleen waargenomen op kleine eilandjes in de rivier die een goed deel van het jaar ontoegankelijk zijn voor dieren. Bladeren, schors en vruchten dienen als voedsel voor veel van de wilde dieren die zich graag tegoed doen aan de jonge bomen waardoor deze niet verder groeien. Wil de boom overleven, dan zullen delen van de uiterwaarden langer onder water moeten blijven zodat ze niet makkelijk toegankelijk zijn voor dieren en zo verder kunnen groeien.

Om de totale verdwijning van *F. Albida* te voorkomen moeten wilde dieren gedurende zeker zes maanden op afstand gehouden worden. Dat kan worden bereikt door een tweeledig reservoir beheer te hanteren: (i) een gematigde overstroming van 5,800 m^3/s eens in de 5 jaar gedurende 5 tot 6 weken in de maanden februari / maart; (ii) het sluiten van de schuiven gedurende een periode van 3 tot 4 weken uit te voeren om er zeker van te zijn dat de oevers ook in de maanden mei/juni nat genoeg zijn voor het proces van ontkiemen. Een geleidelijke sluiting van de schuiven heeft tevens tot gevolg dat het grondwaterpeil niet te snel daalt, waardoor verzakkingen en oevererosie wordt voorkomen.

Table of Content

CHAPTER 1: INTRODUCTION

1.1 General Background

The Zambezi River has its source in the Kalene Hills, 1,450 meters above sea level, in North Western Zambia. The river can be divided into three segments: the Upper, which is 1,078 km long, from the source to the Victoria Falls; the Middle, which is 853 km long, between the Victoria Falls and Cahora Bassa Gorge; and the Lower, which is 593 km long, from Cahora Bassa to the Indian Ocean, giving a total length of 2,524 km from the source to the sea.

The basin of the Zambezi is one of the most diverse and valuable natural resources in Africa and constitutes one of Southern Africa's most important natural resources (IUCN, 1996, Timberlake, 1998, World Bank, 2010). It covers an area of 1,357,000 km^2 and extends over portions of eight Southern African Countries (Figure 1.1). In addition to meeting the basic needs of more than 40 million people and sustaining a rich and diverse natural environment, the Zambezi River plays a central role in the economies of the riparian countries, providing important environmental goods and services which are essential for regional food security and hydropower production (World Bank, 2010). Given the rapid rate of development and population growth in the Southern African region and the existing shortage of water for urban, industrial and agricultural use, many are looking to the Zambezi River for water and hydropower (Timberlake, 1998). Currently, a total of approximately 5 GW is being harnessed from the river using five power stations. A further hydropower potential of 6,370 MW is under development, while 3,750 MW is yet to be developed.

Since the major focus is on the generation of power, relatively little attention is given to environmental considerations (White, 1969, Attwell, 1970, Timberlake, 1998). In particular, there is a lack of comprehensive research in the pre-impoundment phases, which is a common feature of man-made lakes in Africa (Timberlake, 1998). Very little attention is given to the downstream river and its riparian areas, which are affected by the impoundments. This is surely true for the Middle Zambezi, as most of what exists as pre-impoundment information concerns the area to be inundated in terms of fisheries and ecological surveys (Jackson, 2001).

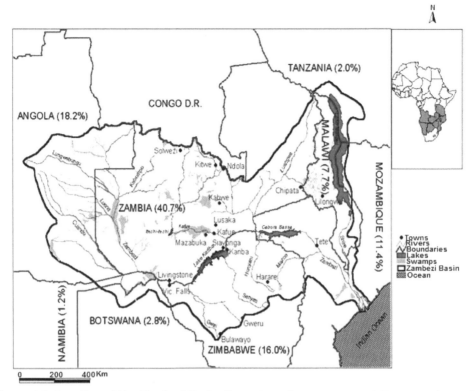

Figure 1.1: Location of the Zambezi Basin. Percentage figures represent the percentage of the Zambezi River Basin within a given country (Source: ZRA, 2000).

The research presented in this thesis focuses on the definition of the environmental flow regime for the Middle Zambezi River reach. The work includes the study of the interaction between surface and subsurface water flows, sediment, river morphodynamics, floodplain vegetation and fauna. The study area is located downstream of two hydropower schemes: Kariba and Kafue, which means that most flows through this river reach are regulated according to hydropower generation needs.

The period (2010 - 2015) in which this research work was undertaken coincided with a critical electricity power deficit of more than 4 GW in the Southern African region (SAPP, 2014), which resulted in widespread and extended interruptions of electricity supply. This made the governments and the power utilities to double the effort in investment and development of new hydropower schemes. In response to finding a solution to the regional power deficit, four hydropower sites, upstream of the study area, have commenced the development process. With this effort, there is now only one site remaining for development (the 1,200 MW Devil's Gorge) upstream of the study area. Table 1.1 lists the existing, those under development, and newly proposed hydropower schemes upstream of the study area.

Table 1.1: Existing, under development and proposed hydropower schemes upstream of the study area.

MAIN EXISTING HYDROPOWER SCHEMES UPSTREAM OF THE STUDY AREA			MAIN HYDROPOWER SCHEMES UNDER DEVELOPMENT UPSTREAM OF THE STUDY AREA			MAIN PROPOSED HYDROPOWER SCHEMES UPSTREAM OF THE STUDY AREA		
No.	Name of Power Station	Installed capacity (MW)	No.	Name of Scheme	Power Potential (MW)	No.	Name of Scheme	Power Potential (MW)
1	Victoria Falls	108	1	Itezhi-tezhi	120	1	Devils Gorge	1,200
2	Kafue Gorge	990	2	Kariba South Extention	300			
3	Kariba South Bank	750	3	Batoka Gorge	2,000			
4	Kariba North Bank	1,080	4	Kafue Gorge Lower	750			
	Total	2,928		Total	3,170		Total	1,200

The study area is a part of the Middle Zambezi sub-catchment. It is an area of immensely rich biodiversity supporting the Lower Zambezi National park on the Zambian side and Mana Pools, Sapi and Chewore National Parks on the Zimbabwean side, which were designated as UNESCO World Heritage Site in 1984.

In its inscription, the following is given as the reason for declaring the area as a World Heritage Site: "The river and the sand-banks that are formed by erosion and deposition are key to the exceptional natural value of these areas. On the riverine strip, large groups of animals congregate annually during the dry season when water elsewhere is scarce. The area is one of the three most important refuges for black rhino in Africa. Also, over 6,500 elephants, 11,000 buffalo, lions, hippos, crocodiles, leopards and cheetahs live in these protected areas. Among other sites it stands out for its good management standards and strictly controlled hunting" (UN, 1992-2016 UNESCO World Heritage Centre website). The consideration of sustainability arises when taking into account the study area with a unique combination of being a rich habitat supporting wildlife, and being an area with such hydropower regulation domination.

1.2   The Hydropower-to-Environment project objectives and desired outcomes

The research work presented in this thesis was undertaken as one of the four key research components of the Hydropower-to-Environment (Power2Flow) project, funded by the UNESCO-IHE Partnership Research Fund (UPaRF). Therefore an overview is given of this wider project, aiming at balancing ecosystem health with hydropower generation in the hydropower dominated Zambezi basin. The Power2Flow project commenced in March 2010 with the lead institution being UNESCO-IHE. The project had the following partner institutions: Eduardo Mondlane University (Mozambique); WaterNet; Eidgenössische Technische Hochschule (ETH) Zurich, Switzerland; WWF the Netherlands; WWF Zambia and WWF Mozambique (UNESCO-IHE, 2009).

In view of hydropower development in the Zambezi Basin, the overall objective of the project was to improve the sustainability of reservoir operation through the development of policy instruments, including reservoir operating policies and cost-sharing mechanisms, which balance water uses for the environment and for energy generation. To achieve this overall goal, four specific objectives, each constituting a scientific research challenge, were outlined:

- Specific objective 1: valuating river ecosystems' goods and services to ultimately derive demand curves for environmental flows at various points in space and time in the Zambezi basin.
- Specific objective 2: to improve the understanding of the complex response of ecosystems to flow regime in large regulated rivers and their floodplains, focusing on flow and biogeomorphological interactions, their links to the ecosystem response and thus goods and services.
- Specific objective 3: to develop a multipurpose multireservoir operation model that can identify reservoir operating policies from a mix of traditional objectives (e.g. hydropower) and environmental objectives (e.g. ecological flows), and using the outputs of specific objectives 1 and 2.
- Specific objective 4: to develop adequate policy instruments and regulatory frameworks to effectively consider environmental flows when dispatching hydropower plants in a loose power pool (self-dispatch hydrothermal electrical system).

The use of the research instruments was identified as the key approach towards making the overall Power2Flow objective measureable and achievable. The research work presented in this thesis serves to meet the project specific objective 2 which aims at assessments of how flows interact with the river and floodplain morphology and riparian vegetation. This research has benefited from five MSc. research findings and data, while the results from this research will also serve as input to the other MSc and PhD studies that will be undertaken on other power2flow specific objectives. This interrelationship is presented in Figure 1.2.

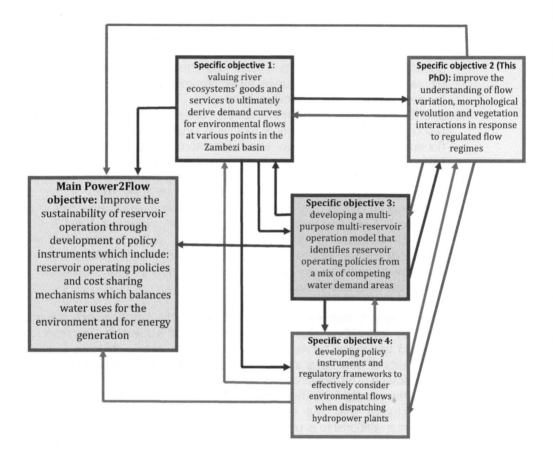

Figure 1.2: Interrelationship of this PhD to the Power2Flow objectives

1.3   Research Objectives

The main objective of the research is to establish the environmental flow regime for the Middle Zambezi reach that minimises the impact of the upstream hydropower schemes on the river environment. This requires the assessment of the effects of regulated water flow on the short and long term morphological evolution of the river, floodplain vegetation and grazing animals, as well as their interactions. The assessment requires the study of the pre-impoundment and post-impoundment states of the Middle Zambezi and future state scenarios.

1.3.1 Key Research questions

To effectively direct the research towards meeting the main objective, the following six research questions should be answered:

Question 1:   What is the current state of the Middle Zambezi River reach and its floodplains in view of the present flow regulation from the Kariba and Kafue hydropower schemes?

Question 2:   What was the state of the river reach and its floodplains before Kariba and Kafue hydropower schemes were constructed?

Question 3:   What is the difference between the past and the current state of the middle Zambezi River reach and floodplains?

Question 4:   What is the role of the tributary streams in discharge and sediment supply to this river reach?

Question 5:   What are the interactions between floodplain vegetation, surface and subsurface flows?

Question 6:   What will be the state of river reach and floodplains in the future if the current water regulation remains the same?

Question 7:   What is the environmental flow regime that is required to minimise the impact of the upstream hydropower schemes?

1.4   Relevance of the research

Why is this research important?

In hydropower dominated basins like the Zambezi, the main focus for the governments and the water resources management institutions is mainly to harness the power potential that flow with the water into the Indian Ocean. With this focus, little or no attention is made to incorporate aspects that would sustain the downstream environment. The lack of attention to certain key aspects at design stage of the dams has resulted in what the World Commission on Dams described as a failure of large storage facilities worldwide to meet their expected performance due to lack of attention devoted to the operational aspects once the storage project is completed (WCD, 2002). The aspect of lack of environmental considerations has contributed to this conclusion of poor performance of large storage facilities, especially with the current global pressure on consideration of environmentally-oriented objectives, such as environmental flow requirements that were historically overlooked in the project development and licensing phases (Brown and Watson, 2007). In recent years, through close collaboration of stakeholders and WWF, effort has been made on the Kafue River, one of the major tributaries of the Zambezi, to improve and modify the dam operating rules for the Itezhi-tezhi and Kafue Gorge Dams by taking into account environmental requirements (DHV, 2004). From this collaborative effort, it has became evident that with research, flood forecasting and simulation tools to support decision making, water resources managers are able to facilitate flow releases for the environment (MEWD, WWF and ZESCO, 2003).

Since hydropower development was the major focus in the development stages of the water resources infrastructure on the Zambezi River, few funds were allocated for research and resources inventory before the construction of the Kariba Dam. The lack of attention to the environment downstream of the dam, gives an impression that there is nothing much in the Middle Zambezi Valley to warrant investment into research and data collection. However, the middle Zambezi floodplains scored quite highly in terms of vegetation, mammals, birds, herps (reptiles and amphibians), and fish species in Timberlake (1998) assessment of wetlands of the Zambezi. The Timberlake ranking was based on the literature that was reviewed on all the Zambezi Basin wetlands. Though the area was assigned a score of 4, which was the highest score to show the level of significance of the wetland resources, the International Union for Conservation of Nature (IUCN) funded study did not even cover this area in a detailed assessment. Because of the rich biodiversity that the Middle Zambezi floodplains support, more attention in terms of research needs to be given to the area. This research aims at contributing to the generation of the much-needed information and data to support the decision making process on allocating funds and investments needed to restore and conserve the area, in order to facilitate the sustainability of the rich biodiversity that makes the habitat so special.

Why is integrating water, sediments, floodplain vegetation and animals important for the definition of environmental flow regime?

Environmental flow regime is not only the minimum flow that has to be allocated to a river, it is a series of discharges, including low and high water flows, minimising the impact of flow regulation. When the flow discharge in a river is changed, the entire system is altered (Crosato, 2014, Lu and Siew, 2006, Richter and Thomas 2007, Bunn and Arthington, 2002, Matos et al., 2010). Richter and Thomas (2007) argue that the alteration of natural water flow regimes wrought by dam construction and operation has had the most pervasive and damaging effects on river systems. The river system includes abiotic (physical environment made of water and sediments) and biotic (flora and fauna - both aquatic and terrestrial) components. These components of the river system interact and influence each other at all spatial and temporal scales. In this regard, it is not possible to define a sustainable management solution without considering and understanding the interactions between the different components. It is this understanding that allows identifying the impacts of flow regime alterations and how restoration and sustainability of the system can be achieved. Failure to understand the interactions would lead to environmental flow prescriptions that do not foster sustainability. Therefore, this research focuses on the assessment of these critical interactions to determine what can be done to miminise the impact of flow regulation and define an optimum environmental flow regime for the Middle Zambezi River.

1.5   General approach

Although the existence of the Middle Zambezi floodplains can be attributed to geological processes that led to the formation of a valley trough, yet the floodplain features are a creation of the hydromorphological processes related to the interaction of flows, morphology and vegetation. To bring out a good understanding of the interaction of these features, a combination of research methodologies were employed.

The approach in this research is a mixture of the following: a complete literature review covering the processes involved and the situation of the Middle Zambezi River in the present and in the past; field-based observations and measurements; laboratory study of plant growth; historical map and satellite image analyses; hydrodynamic and morphodynamic modelling (Figure 1.3). In order to achieve the overall objective of the study, the research was divided into six components, each one having a specific approach and methodology. Some components were elaborated into five MSc research projects carried out in UNESCO-IHE and University of Zimbabwe. Figure 1.4 shows the layout of the research components, showing the MSc. Research components that contributed to this PhD research. The MSc. research components were as follows:

- MSc. 1 - The interaction of flow regime and the terrestrial ecology of the Mana Floodplains in the middle Zambezi River Basin (Ncube, 2011);
- MSc. 2 - Hydrological analysis of the Middle Zambezi and impacts of the hydropower dams on the flow regime in Mana Pools National Park (Ekandjo, 2011);
- MSc. 3 - Effects of flow alteration on Faidherbia albida stands of the Middle Zambezi Floodplains (Gope, 2012);
- MSc. 4 - Effects of large Dams on Riverine Geomorphology and Riparian vegetation Case Study of Mana Pools Floodplains, Middle Zambezi (Mubambe, 2012; and
- MSc. 5 - Effects of dam operations on the Mana Floodplain forest in the Middle Zambezi - the role of flow regime, river morphology and groundwater (Khan, 2013).

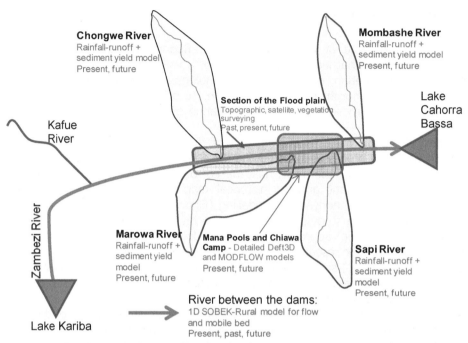

Chongwe River
Rainfall-runoff +
sediment yield model
Present, future

Mombashe River
Rainfall-runoff +
sediment yield model
Present, future

Lake
Cahorra
Bassa

Section of the Flood plain
Topographic, satellite, vegetation
surveying
Past, present, future

Kafue
River

Zambezi River

Marowa River
Rainfall-runoff +
sediment yield
model
Present, future

**Mana Pools and Chiawa
Camp** - Detailed Deft3D
and MODFLOW models
Present, future

Sapi River
Rainfall-runoff +
sediment yield
model
Present, future

River between the dams:
1D SOBEK-Rural model for flow
and mobile bed
Present, past, future

Lake Kariba

Figure 1.3: Combination of research methodologies.

Observation and analysis of water flows should cover short and long temporal scales. Daily flows are needed to observe the diurnal variation and the interaction with groundwater flow through riverbanks. Annual and seasonal flows are needed to identify key interactions between floodplain vegetation and the river. Decadal and long-term flows are needed to study the river morphological changes. The approach also includes numerical modelling of surface and ground water flows, as well as morphodynamic changes.

Different numerical approaches were used to study the different components. The groundwater simulations are meant to assess the ground water flow direction throughout the year. The morphodynamic and hydrodynamic simulations, calibrated on the past evolutions, are meant to study scenarios and to assess the effects of regulated regimes on floodplain floods and river morphology. Three different models are selected: the open-source Delft3D (www.Deltares.nl), for the study of the river morphodynamic evolution in two dimensions (2D); the 1D software package SOBEK-RURAL (Deltares, 2011), for the study of the hydrodynamic scenarios and floodplain inundation; MODFLOW (Panday et al., 2013; Zhou, 2012), for the groundwater flow analysis.

Field surveys were carried out to complement lacking data. In particular hydrodynamic, ground water and sediment data are lacking and require continuous measurements covering at least one year. This means that instruments have to be placed at selected locations in the river and in wells excavated in the floodplain.

The *Faidherbia albida* tree has been identified as the biological indicator for the floodplain vegetation. Field surveys are required to study the distribution of the tree on the floodplain, the age of the trees on the floodplain and interaction between the tree, the floodplain flow and the grazing animals during all tree development stages. Furthermore laboratory investigations are needed to assess the growth rates of the roots and the tree stem after germination. A thorough integration of all the findings is finally needed to be able to identify cause and effects and draw a conclusion on possible mitigations. The environmental flow regime arises from this final analysis.

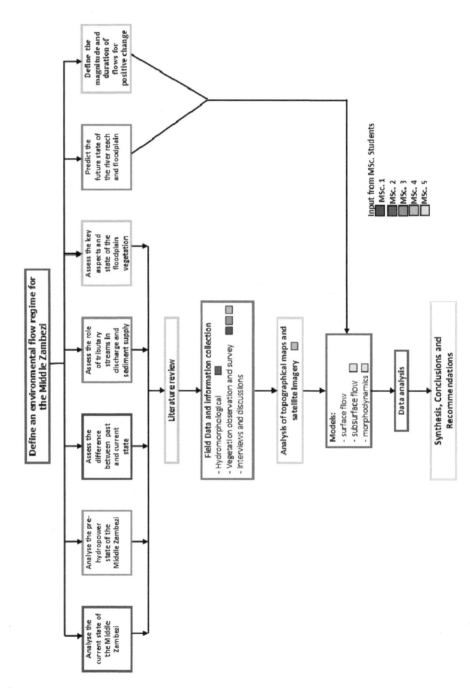

Figure 1.4: Layout of research components.

11

1.6   Structure of the thesis

This thesis is organized in eight chapters.

Chapter 1 presents the introduction by highlighting the general background to the research and goes further to give a brief on the objectives of the Power2Flow project under which this PhD was financed and undertaken. The rational of the research, the objectives and the general approach are also given in this chapter.

Chapter 2 presents a historical perspective of the hydropower dominated Zambezi Basin by briefly describing the key drivers of the hydropower development and further gives the hydrology and the hydropower potential setting of the basin. The chapter further goes into details to describe the current operating hydropower schemes and those under development. A brief is then given on the basin-wide regional initiatives for hydropower operation collaboration as the platform on which the research findings presented in this thesis can be shared and discussed.

Chapter 3 is dedicated to the detailed presentation of the study area, the Middle Zambezi sub-catchment, giving a general layout and location, as well as a specific brief on the socio-economic, physical and ecohydrological features. In outlining the ecohydrological setup, floodplain vegetation, the fauna and key observed threats are outlined.

Chapter 4 deals with the sub-catchment surface and subsurface flows. In particular, giving the hydrological set up, the various field measurements and analyses, the surface water balance, the surface and subsurface flow model simulations are illustrated. The chapter concludes with the analysis of the research results.

Chapter 5 focuses on the morphodynamics of the Middle Zambezi reach, presenting the results of sediment collection and analysis, the historical morphological changes using historical maps and satellite images. Further, the chapter outlines the morphodynamic model simulations and concludes by presenting the research results.

Chapter 6 is dedicated to floodplain vegetation with focus on the choice of the biological indicator, its nature and distribution. The field and laboratory experiments are also presented. The floodplain forest distribution derived from satellite image analysis is also described. The chapter concludes with the research results on the tree characteristics, age distribution and early stages of tree growth and discusses the implications on how the sustainability of the floodplain forest interacts with floodplain flows.

Chapter 7 deals with the derivation of the environmental flow regime for the Middle Zambezi. The chapter starts with a literature review on previous methods to derive the environmental flow regime. Different scenarios of water management are analysed using the 1D hydrodynamic model and the short-term and the long-term effects of the suggested environmental flow regime are presented. A conclusion is then given on the definition of the environmental flow regime for the Middle Zambezi

The last chapter of this thesis is Chapter 8. It presents a synthesis of this research, discusses the results and lists the conclusions and recommendations.

CHAPTER 2: HISTORICAL PERSPECTIVE OF THE HYDROPOWER DOMINATED ZAMBEZI BASIN

2.1 Driver of hydropower development

2.1.1 Availability of favorable geological formation

Hydropower is a reliable and well-understood form of renewable energy production. Globally, hydropower is said to produce more renewable electricity than any other technology (Kumar et al., 2011). There are many suitable locations for hydropower schemes, the main requirements being the availability of flowing water (Discharge) and height difference between the water intake point and the generation point (Head). Because of these two naturally occurring physical features, discharge and head, not every river can produce hydropower, therefore, the moment the head is observed in a river profile, the interest of harnessing hydropower arises (Kumar et al., 2011).

The key hydropower variables of Q (Discharge) and H (Head) are seen in the hydropower equation (2.1).

$$P(kW) = \frac{\eta \gamma Q \left(m^3 / s \right) H \left(m \right)}{1000} \qquad (2.1)$$

Where:

P = Generation output in (kW)

Q = Water flow through the turbine (Discharge) in (m^3/s)

H = Net head of water (m) - the difference in water level between upstream and downstream of the turbine

η = Station Efficiency

γ = Specific weight of water (N/m^3).

The key driving factor that makes the Zambezi River attractive for hydropower development lies in its favourable longitudinal profile, presenting a number of locations where prominent head is observed. The Kafue river, a tributary of the Zambezi in its Middle course, has the spectacular head of 600 meters in a distance of 25 kilometers. Figure 2.1 shows the longitudinal profiles of the Zambezi and Kafue Rivers.

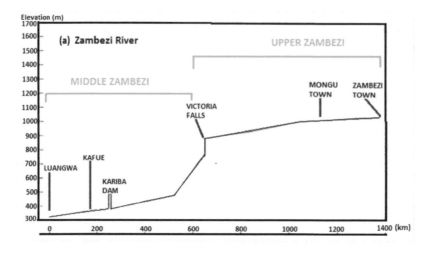

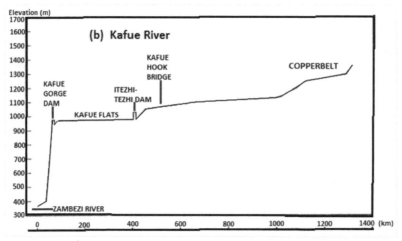

Figure 2.1: Longitudinal profiles for Zambezi(a) and Kafue(b) Rivers (adapted from Yachiyo, 1995).

2.1.2 Industrial development energy demand

For all the existing hydropower schemes in the Zambezi Basin, the main driver is the need for energy to drive the industries, mainly targeting the copper mines and other mining developments in Southern Africa. For Cahora Bassa, located at the downstream end of the Middle Zambezi reach, the main target and market is the industrial development market in South Africa. The critical need to meet the energy demand for the industries is highlighted in the advertisement which was placed by Standard Bank in the book by Malvern(1960), the advertisement was as follows: "Progress is symbolised in the pylons which are being erected all over the Federation. Essential to us all - Kariba's electrical power will be the life blood of our growing industries and is to be carried in these pylon power lines which run like arteries to the most distant parts of our country..." It is clear that in the Nineteen Sixties, the political leaders were looking for sustainable and cheaper renewable

15

energy to power the industries, and the hydropower potential which Zambezi carried was the target (Malvern 1960). Projections on how much coal was going to be saved made the hydropower projects very attractive, which was used as a justification for the investment. An example of the projection on how much coal was going be to be saved by switching to Kariba hydropower, is as shown in Figure 2.2.

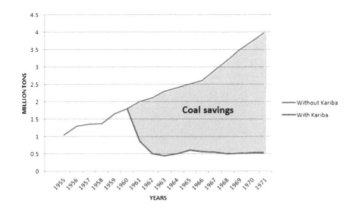

Figure 2.2: Projected coal savings with the switch to hydroelectric power from Kariba (adapted from Malvern, 1960).

2.2   Hydrology and hydropower potential set up

2.2.1 Hydrology

The basin of the Zambezi River receives a mean annual rainfall of about 950 mm, this ranges from the high rainfall in the north of 1,400 mm and lower rainfall in the south of 600 mm. Most of the rainfall is concentrated within the summer period of October – April. The north and east parts of the basin experience significantly more precipitation than the south and west parts. Less than 10% of the mean annual rainfall in the basin flows through the Zambezi River to the Indian Ocean. The mean annual discharge at the outlet of the Zambezi River is 4,134 m^3/s. The Zambezi River Basin is characterized by extreme climatic variability, therefore the river and its tributaries are often subject to a cycle of floods and droughts that have devastating effects.

To facilitate calculation of runoff from such rainfall variable catchment, the basin has been subdivided in 13 sub-basins, as presented in Figure 2.3 and the hydrological variables have been calculated for each sub-basin as presented in Table 2.1.

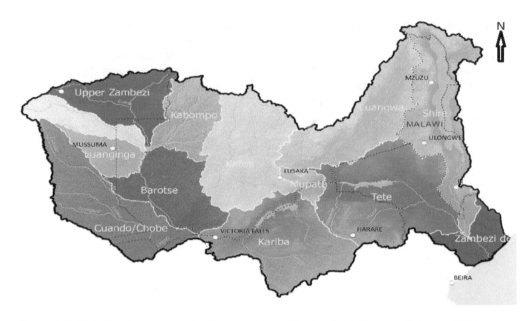

Figure 2.3: Main Zambezi sub-basins (Adapted from Map by J.-M. Mwenge Kahinda, CSIR-South
    Africa in Beilfuss, 2012).

Table 2.1: Zambezi sub-basins hydrological variables (World Bank, 2010 and Beilfuss, 2012).

Sub-catchment	Area (Km^2)	Hydrological variables				
		Mean Annual Rainfall (mm)	Potential Evapo-transpiration (mm)	Runoff Coefficient	Mean Annual runoff (Mm^3)	Cumulative Zambezi mean annual runoff (Mm^3)
UPPER ZAMBEZI						
Upper Zambezi	91,317	1,225	1,410	0.21	23,411	23,411
Kabompo	78,683	1,211	1,337	0.09	8,615	32,026
Lunguebungo	44,368	1,103	1,472	0.07	3,587	35,613
Luanginga	35,893	958	1,666	0.06	2,189	37,802
Cuando/Chobe	148,994	797	1,603	-	0	37,802
Barotse	115,753	810	1,578	-	-553	37,249
MIDDLE ZAMBEZI						
Kariba	172,527	701	1,523	0.05	6,490	43,712
Kafue	155,805	1,042	1,780	0.07	11,734	
Mupata	23,483	813	1,708	0.09	1,703	57,127
Luangwa	159,615	1,021	1,555	0.10	16,329	
LOWER ZAMBEZI						
Tete	200,894	887	1,436	0.10	18,007	91,463
Shire River/Lake Malawi	149,159	1,125	1,643	0.09	15,705	
Zambezi Delta	33,506	1,060	1,652	0.10	3,564	110,732
ZAMBEZI BASIN	**1,409,997**	**956**	**1,560**	**0.08**		**110,732**

2.2.2 Hydropower potential

The Middle Zambezi has substantial hydropower potential, with new hydropower projects totaling more than 5,000 MW in various stages of consideration. Potential new power generation schemes include the 2,000 MW Batoka Gorge, 1,200 MW Devils Gorge, and 640 MW Mupata Gorge dams on the mainstream Zambezi, and the 750 MW Kafue Gorge Lower amd 120 MW Itezhi-tezhi on the Kafue River. Proposed extension to existing power stations would increase power output by about 300 MW at Kariba South Bank Power Station. It is estimated that the total hydropower potential on the Zambezi can amount up to 15,000 MW (Word Bank, 2010, Beilfuss, 2012). Figure 2.4 shows a schematic river profile with the existing hydropower schemes and those under development at various stages on the main Zambezi and the Kafue tributary.

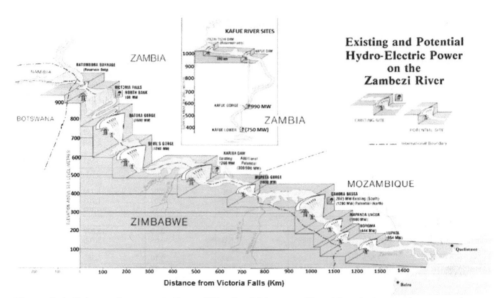

Figure 2.4: Schematic presentation of Zambezi River profile with existing and proposed hydropower sites (ZRA, 2011).

2.3   Developed hydropower schemes, and operating rules

The Zambezi Basin system of hydropower schemes can be well explained in the schematic outline in Figure 2.5 which show the main existing schemes on both the main stream and the tributaries. Each hydropower scheme has specific operating rules that guide water regulation in the water reservoirs.

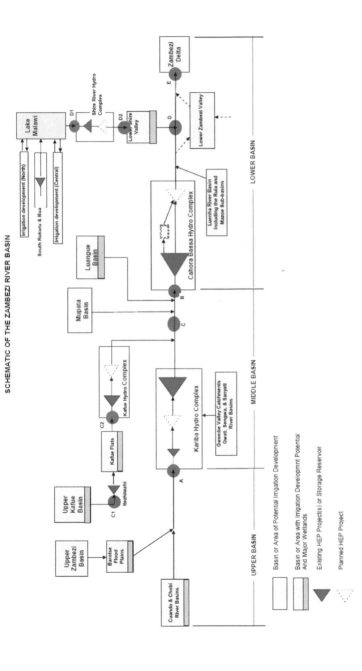

Figure 2.5: Schematic outline of the existing and proposed hydropower schemes of the Zambezi Basin (ZRA, 2011).

2.3.1 Victoria Falls Scheme

The Victoria Falls hydropower scheme has an installed capacity of 108 MW and is a run of the river hydro power plant. The Power station is located on the left back, downstream of the Victoria Falls. The water diversion intake channel is located some 500 meters upstream of the falls. Refer to Figure 2.6 showing the location of the Power station relative to the Victoria Falls.

Figure 2.6: Location of the Victoria Falls Power station downstream of the Victoria Falls.

The Victoria Falls power station was built in three phases. The first, Station A (8 MW), was commissioned in 1938 while the second, Station B (60 MW), was built underground, giving an additional capacity, in 1969. The third, Station C (40 MW), was completed in 1972 providing an installed capacity of 108 MW. The Station is run using a diversion tunnel with the following water requirements: Station (A) a discharge of 10.5 m3/s; Stations B and C a discharge of 106.7 m^3/s. The water right has an environmental requirement to reduce water abstraction when the Zambezi River flows reduce to 400 m^3/s. This condition is due to the sensitivity of the station location and the need to maintain the esthetic beauty of the falls. The hydrograph and the limit of 400 m^3/s for reduction in water abstraction is as shown in the hydrograph in Figure 2.7.

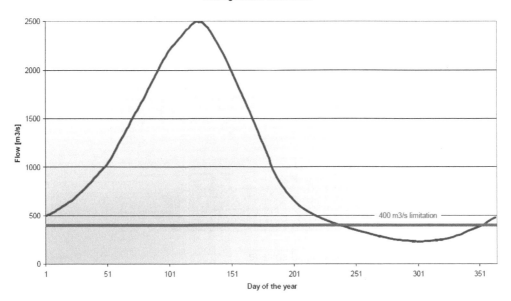

Figure 2.7: Average flows and lower flow limitation for abstraction for generation at Victoria Falls.

2.3.2 Kariba Scheme

The Kariba scheme comprises two power stations: the Kariba South Bank, commissioned in 1958 with an installed capacity of 750 MW, and the Kariba North Bank, on the right bank, commissioned in 1976 with the current installed capacity of 1020 MW (after commissioning of 360 MW Kariba North Bank Extension). Refer to Figures 2.4 and 2.5 for relative location of Kariba in the Zambezi. Lake Kariba provides the water storage to meet the generation needs of the power stations, Figure 2.8 shows the Kariba Dam with two spillway gates open. The Kariba reservoir characteristics are as outlined in Table 2.2. The two power stations are given an annual allocation of the amount of water to use and the power stations' water abstraction is monitored by the Zambezi River Authority who also manage and regulate the Kariba reservoir. The reservoir operating rules follows the reservoir level lower rule curve. The Kariba Dam lower rule curve is as shown in Figure 2.9.

Figure 2.8: The Kariba Dam and flood spillage.

Table 2.2: Kariba Reservoir Characteristics (Source: ZRA, 2011).

Elevation	Area	Volume	Outflow
m.a.s.l	Km^2	Mm^3	m^3/s
475.5=MOL	**4354**	**54**	-
476	4405	2272	-
478	4608	11278	-
480	4811	20613	7751
482	4991	30408	8168
484	5171	40568	8584
486	5350	51088	8974
488	5531	61998	9347
488.5=FSL	**5577**	**64798**	**9445**

MOL - Minimum Operating Level
FSL - Full Supply Level
masl - Meters Above Seal Level

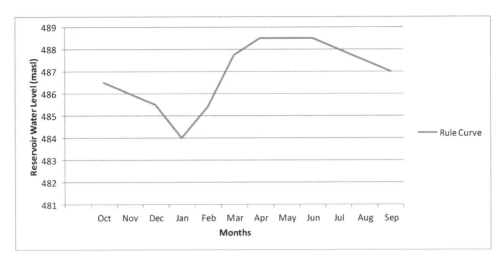

Figure 2.9: Kariba Reservoir Operating Lower Rule Curve.

2.3.3 Kafue Gorge Scheme

The Kafue Gorge scheme has the main water storage reservoir 250 km upstream of the Kafue Gorge at Itezhi-tezhi. The Kafue hydropower scheme has the Kafue Flats in between the Itezhi-tezhi reservoir (upstream) and the Kafue Gorge reservoir (downstream). The Kafue gorge reservoir inflow is therefore controlled by the storage characteristics of the Kafue Flats and results in a 2-month travel time for flow between Itezhi-tezhi and Kafue Gorge dams. The Kafue Flats are a wetland of outstanding significance, serving as a rich habitat to wildlife and as grazing ground for livestock. Due to the availability of water, the Kafue Flats are also important for irrigation agriculture, hosting extensive sugar estates. Refer to the map in Figure 2.10 showing location of the Itezhi-tezhi Dam, the Kafue Flats and the Kafue Gorge dam while Figure 2.11 shows the Kafue Gorge power station generator hall. Figures 2.4 and 2.5 shows the relative location of the Kafue Gorge scheme in the Zambezi basin.

To meet the water needs of both wildlife habitat and agriculture in the Kafue Flats, the Water Right held by ZESCO for the abstraction of water from the Itezhi-tezhi dam is subject to the following three specified environmental flow conditions:

1) The holder shall store and release from the Itezhi-tezhi dam a minimum of 300 m^3/s over a period of four weeks in each year to preserve the ecological balance of the Kafue Flats. This condition is referred to as the March Freshet.
2) The holder shall store and release sufficient water to ensure that a minimum of 15 m^3/s is available for other users between Itezhi-tezhi dam and Kafue Gorge dam at all times.
3) The holder shall ensure that a minimum flow of 25 m^3/s in the river between Itezhi-tezhi and Kafue Gorge dam is maintained at all times.

However, condition 1) has not been respected on a regular basis due to concerns over the availability of water, especially during the drought periods and during the years when there is delay in commencement of the rainy season.

The power station has an installed capacity of 990 MW. The reservoir characteristics of the Itezhi-tezhi and the Kafue Gorge reservoirs are presented in Table 2.4 while the power station characteristics are presented in Table 2.5. The dam operating rules are in accordance to the rule curves shown in Figure 2.12 for Itezhi-tezhi dam and Figure 2.13 for Kafue Gorge dam

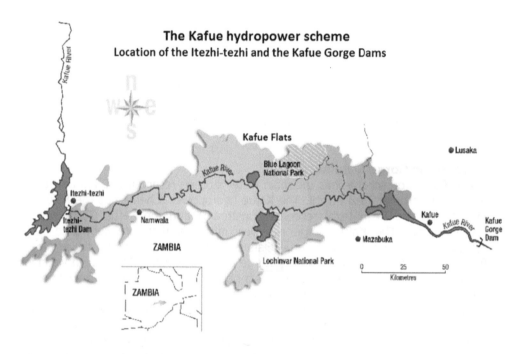

Figure 2.10: Location of the Ithezhi-tezhi and Kafue Gorge Dams (adapted from WWF, 2003).

Figure 2.11: Kafue Gorge Power Station - generator hall (ZESCO, unpublished material).

Table 2.4: Kafue River Basin Reservoir Characteristics

Itezhi-tezhi				Kafue Gorge r			
Elevation m.a.s.l	Area km²	Volume Mm³	Outflow m³/s	Elevation m.a.s.l.	Area km²	Volume Mm³	Outflow m³/s
1006=MOL	90	699	300	970	0	0	308
1009	113	1003	300	972.3=MOL	20	0.01	876
1012	138	1377	300	974	70	69	1420
1015	167	1836	300	975	142	170	1804
1018	203	2387	300	976	430	423	2220
1021	238	3045	525	976.6	805	785	2496
1024	284	3551	1125	977=FSL	1175	1178	2668
1028	346	4746	2355	978	2160	2845	3132
1030.5=FSL	390	6013	4435				
1032	420	6616	5200				

MOL - Minimum Operating Level
FSL - Full Supply Level
Masl - Meters Above Seal Level
Source: ZESCO (unpublished material)

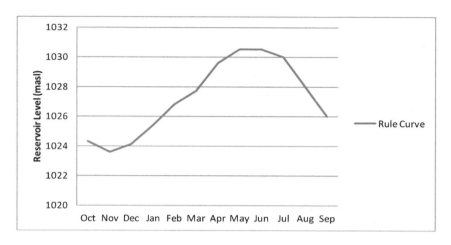

Figure 2.12: Itezhi-tezhi Reservoir Operating Rule Curve (End of the month water level)

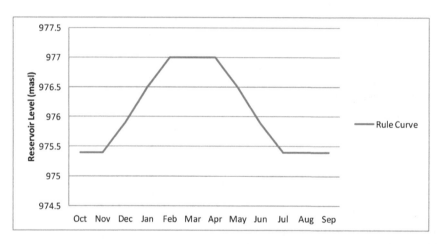

Figure 2.13: Kafue Gorge Reservoir Operating Rule Curve (end of the month water level).

Table 2.5: Kafue Gorge Power Station Parameters (Source: ZESCO, unpublished material).

PARAMETERS	KAFUE GORGE POWER STATION
Installed Capacity [MW]	990
Efficiency	0.88
Head Loss [m]	5
Head (m)	400
Penstock Capacity [m³/s]	290

2.3.4 Cahora Bassa Scheme

Cahora Bassa is managed by Hidroeléctrica de Cahora Bassa (HCB) and has full installed capacity of 2,075 MW. The Cahora Bassa represents the largest hydroelectric power scheme in Southern Africa with the powerhouse containing five 415 megawatts (557,000 hp) turbines. The dam is 171 m high forming a lake, which is 250 km long and 38 km wide, covering a flooded area of 2,700 km^2. This is the fourth largest artificial lake on the African continent. The Cahora Bassa Dam regulates the runoff from the Middle Zambezi catchment between the Kariba and Cahora Bassa Gorges. Figure 2.14 show the dam wall and part of the reservoir. Refer to Figures 2.4 and 2.5 for relative location of the Cahora Bassa Scheme in the Zambezi Basin.

The releases from the Cahora Bassa Dam are governed by hydropower generation requirements and a flood rule curve whereby the reservoir water levels are drawn down prior to each rainy season to provide additional capacity for safely storing and passing the design flood. Spillway discharges are based on all eight gates fully opened, with the crest gate operating for reservoir elevations above 327 m. Minimum water releases for social or environmental purposes are not considered in the rule curve. The existing operating rule for Cahora Bassa is shown in Figures 2.15.

Figure 2.14: Cahora Bassa Dam and reservoir.

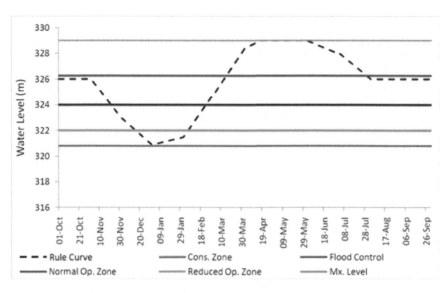

Figure 2.15: Cahora Bassa Operational Rule Curve with different rules at different water levels

(HCB, unpublished material).

2.4 Hydropower schemes under development

The past decade has coincided with a Southern Africa power deficit which has led to the acceleration of hydropower development in the Zambezi Basin. Table 2.6 outlines the various projects under development amounting to a total production of 8,170 MW. These projects are at different phases of development ranging from feasibility to construction phases. Figure 2.16 shows the Itezhi-tezhi project under construction. Refer to Figures 2.4 and 2.5 for relative location of the proposed hydropower schemes in the Zambezi.

Table 2.6: Zambezi Schemes under development.

MAIN PROPOSED HYDROPOWER SCHEMES		
No.	Name of Scheme	Power Potential(MW)
1	Itezhi-tezhi	120
2	Kariba South Extension	300
3	Cahora Bassa Extension	1,200
4	Kafue Gorge Lower	750
5	Batoka Gorge	2,000
6	Devils Gorge	1,200
7	Mupata Gorge	600
8	Mepanda Nkua	2,000
	Total	**8,170**

Figure 2.16: Itezhi-tezhi hydropower plant under construction (ITPC/ZESCO, unpublished material).

2.5 Regional initiatives for collaboration in hydropower scheme operations

River basins of international nature do require transboundary governance solutions to manage catchment issues that cut across the country boundaries. The hydropower related water management as regards to environmental flow regimes do require a basin-wide perspective and a platform of shared interest. This section presents the regional initiatives that exist in looking at implementation of integrated Water Resources Management issues. These initiatives provide the best platform on which research work such as the one presented in this thesis should be discussed for basin wide solutions towards achieving catchment sustainability.

2.5.1 The Zambezi Watercourse Commission
The Zambezi River Basin has been the subject of a long history of sustained efforts to foster regional cooperative development. The evolution of international cooperation in the Zambezi River Basin has taken place over more than three decades, starting in the 1980s, building on the earlier foundations established during the political Federation of Nyasaland and Rhodesia and the development of the Kariba Dam hydropower complex. During the course of efforts to find collaboration, SADC Protocol on Shared Watercourse Systems was developed, was revised in 2000 and ratified in 2003. The conclusion of the protocol on shared water courses is said to have led to the commencement of fresh negotiations in the ZAMCOM Agreement in 2002.

The "Agreement on the Establishment of the Zambezi Watercourse Commission" (the ZAMCOM Agreement) was subsequently signed on July 13, 2004 in Kasane,

Botswana, by the Ministers responsible for water from the majority of the eight Riparian States. The ZAMCOM Agreement came into force on June 19, 2011 after six of the eight Riparian States completed their ratification processes and deposited their ratification instruments with the SADC (2004).

ZAMCOM's goal is to assist the Riparian States achieve regional cooperation and integration through sharing treasured benefits from the water resources of the Zambezi River basin. This is in recognition of the contribution that such cooperation could make towards peace and prosperity in the basin and the Southern African region as a whole.

Associated key influencing factors behind the Agreement include the recognition and consciousness by the Riparian States of the following:

- The scarcity and the value of water resources in the Southern African region and the need to provide the people in the region with access to sufficient and safe water supplies;
- The significance of the Zambezi Watercourse as a major water source in the region, as well as the need to conserve, protect and sustainably utilize its resources;
- The commitment to the realization of the principles of equitable and reasonable utilization as well as the efficient management and sustainable development and management of the basin's water resources;
- The desire to extend and consolidate existing relations of good neighbourliness and cooperation amongst the Zambezi Riparian States on the basis of existing international water instruments.

As stipulated in the ZAMCOM agreement (SADC, 2004), the Commission has the following functions:

   (a) collect, evaluate and disseminate all data and information on the Zambezi Watercourse as may be necessary for the implementation of the ZAMCOM Agreement;

   (b) promote, support, coordinate and harmonise the management and development of the water resources of the Zambezi Watercourse;

   (c) advise Member States on the planning, management, utilization, development, protection and conservation of the Zambezi Watercourse as well as on the role and position of the public with regard to such activities and the possible impact thereof on social and cultural heritage matters;

   (d) advise Member States on measures necessary for the avoidance of disputes and assist in the resolution of conflicts among Member States with regard to the planning, management, utilization, development, protection and conservation of the Zambezi Watercourse;

   (e) foster greater awareness among the inhabitants of the Zambezi Watercourse of the equitable and reasonable utilization and the efficient management and sustainable development of the resources of the Zambezi Watercourse;

   (f) co-operate with the institutions of SADC as well as other international and national organisations where necessary;

   (g) promote and assist in the harmonization of national water policies and legislative measures;

(h) carry out such other functions and responsibilities as the Member States may assign from time to time; and,

(i) promote the application and development of the ZAMCOM Agreement according to its objective and the principles referred to under Article 12 (SADC, 2004).

The ZAMCOM governance has three main organs. These include the Council of Ministers which is the decision making arm; the Technical Committee (ZAMTEC), a technical advisory body; and the Secretariat (ZAMSEC) for overall management and is supported by Project Implementation Unit and Working Groups, refer to Figure 2.17.

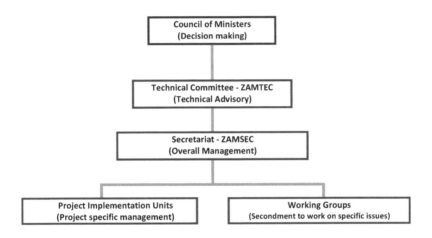

Figure 2.17: ZAMCOM Governance Structure (SADC, 2004).

2.5.2 The Zambezi Water Resources Managers and Dam Operators Committee

The history to regional hydropower related collaboration on the Zambezi basin dates back to the 1950s before the construction of the dams when a Technical Committee was formed. The members included the Federation of Rhodesias and Nyasaland (now Zambia, Zimbabwe and Malawi), Mozambique and South Africa. The main task was to prepare a development plan for the Zambezi Basin. The work of this Technical Committee led to the construction of the Kariba and Cahora Bassa Dams. After the construction of the two dams, the Technical Committee evolved into the current Joint Operations Technical Committee (JOTC) whose membership comprises the Zambezi Basin Water Managers and Dam Operators. Over the years, the JOTC proved to be an effective platform for collaboration for the member institutions responsible for water resources management and dam operations. The current members of the JOTC are from three riparian countries: Mozambique, Zambia and Zimbabwe.

The JOTC has been operating without any formal agreement, but two important agreements were signed on 7th July 2011 and these are: The Memoradum of

Understanding between the governments of Mozambique, Zambia and Zimbabwe; and the agreement among the Zambezi Water Resources Managers and Dam Operators leading to the official creation of the JOTC.

The key aim of the JOTC is to facilitate collaboration and exchange of hydrometeorological data and dam information for efficient utilisation and management of water resources in the catchment areas of the Zambezi water course amoung the JOTC member institutions and other stakeholders.

The Zambezi Water Resources Managers and Dam Operators have proposed a structure for governance as outlined in Figure 2.18.

Figure 2.18: JOTC Governance structure (ZRA et al., 2011).

CHAPTER 3: THE STUDY AREA - MIDDLE ZAMBEZI SUB-CATCHMENT

3.1 General description and location

The Study area falls within the Middle Zambezi, being downstream of the Kariba Dam to the Mupata Gorge, upstream of the Cahora Bassa Dam. The specific location of the study area is bound by latitude 17^0 30'S to the south and latitude 15^0 07'S to the north, with longitude 30^0 15'E to the east and longitude to the 28^0 00'E west. Figure 3.1 shows the location of the Zambezi Basin relative to Southern Africa and shows the location of the Middle Zambezi Study area. The Middle Zambezi is a host to three main hydropower schemes with installed generation capacity of about 3 GW and five hydropower potential sites with a total of about 4 GW capacity. The study area is located downstream of the all the existing hydropower schemes in the Middle Zambezi. Therefore this area can be described as a hydropower dominated subcatchment with most of the river flows being hydropower regulated from the existing reservoirs of Kariba, Ithezi-tezhi and Kafue Gorge. In the mean time, the Middle Zambezi is important for its ecology including the Mana Pools National Park and the Lower Zambezi National Park.

The middle Zambezi and its associated floodplains start at about 88 km downstream of Kariba Dam and is approximately 70 km from the Cahorra Bassa Lake as shown in Figure 3.2 colour coded in shade of blue for elevation. The total length of the flood plain is 91 km. The widest part of the flood plain is where the Chikwenya island is located, with two river confluences both on the right and left bank, it is 7 km. Whereas the width of most of the sections is between 3.5 and 5 km. The narrowest point is the Vundu point, upstream of the Mana pools and it is where the floodplain is only about 1km wide. The floodplain in this study is described as the valley floor adjacent to the river channel which becomes inundated at high flows. The benefits of floods in this area include: replenishment of top soil and nutrients supplies on the flood plain, providing water to seedlings and trees requiring periodic inundation, allowing aquatic animals to migrate to calmer, nutrient-rich shallows in the flood plains to feed and breed (Gordon et al., 1992).

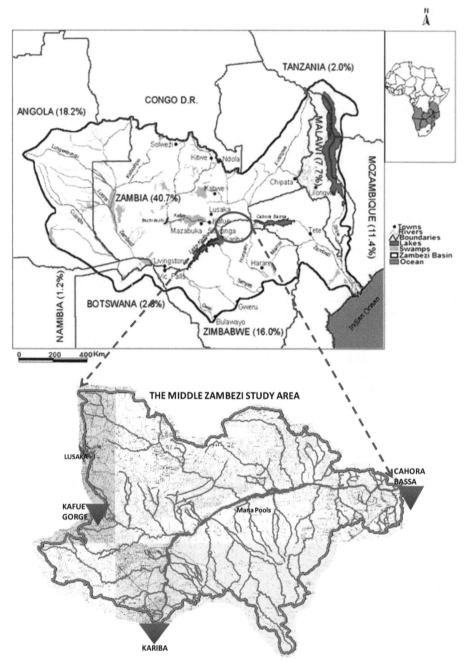

Figure 3.1: Location map of the Middle Zambezi Study area (Map of Zambezi Basin - ZRA and study area map developed from1:50,000 scale topographical maps).

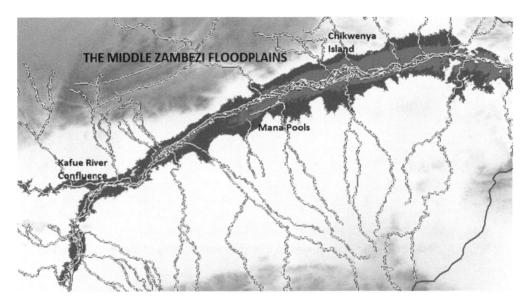

Figure 3.2: Middle Zambezi Floodplains in shade of blue range from 331 - 366 meters above sea

level (Developed from 90m Digital Elevation Model data from SRTM_42_16).

3.2 Socio-economic features

3.2.1 Wildlife management economic activities

About 70 percent of the study area comprise the wildlife management parks. On the Zambian side they are: the Chiawa Game Management Area; the Lower Zambezi National Park; and the Rufunsa Game Management Area. On the Zimbabwean side they are: the Hurungwe Safari Area; the Mana Pools National Park; the Sapi Safari Park; the Chewore Safari Area; and the Dande Safari Area. Figure 3.3 shows the wildlife management areas on both the right and left bank of the Zambezi, and show some large mammals found in the area.

Apart from the hydropower generation, the key socio-economic activity in the study area is wildlife management and associated tourism activities for both the Zambian and Zimbabwean sides. Both the left and right banks of the river have tourist lodges and these are also associated with provision of employment to the local people. Figure 3.4 shows the Chiawa Camp Lodge on the left bank and the Mana Pools Lodge on the right bank.

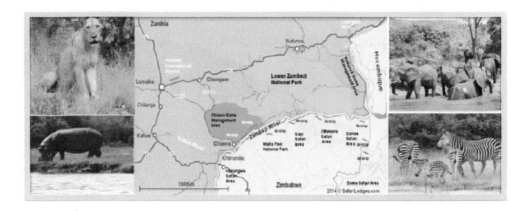

Figure 3.3: Map of wildlife management parks and some animals of the Middle Zambezi.

Figure 3.4: Types of lodges in the Wildlife Management Parks (Chiawa Camp and Mana Pools Lodges).

3.2.2 Agricultural related activities

The other activity of economic importance is irrigational agriculture. Centre pivot and sprinkler irrigational schemes are commonly used. The agricultural activities are mainly on the Zambian side in and outside the Game Management areas. The main crops grown are bananas, wheat and maize. The development of irrigation agriculture in this area is attracted by the constant abundant water available throughout the year. The farmers in this area being downstream of the hydropower schemes have no water restrictions. Therefore for the commercial farms by ZAMBEEF and Chakanaka, the only limiting factor for their expansion is land availability, because water is abundant. Figure 3.5 shows the ZAMBEEF Farms.

Figure 3.5: Commercial farming activities in the Study area (ZAMBEEF).

At subsistence level, the local people in the Chiawa Game Management area under Chietainess Chiawa have historically been dependant on the floodplain and islands for recession agricultural practices. As gathered from the informal discussion with the local people (see Figure 3.6), though, the flooding system has changed dramatically, the local people still take chances to grow crops in flooding zone hoping that the flood gates are not opened before they harvest. In most cases, since the notification of the opening of flood gates is given at a short notice, the local people suffer from flooding of crops in the seasons when Kariba does open the gates. Despite the negative consequences and risk associated to growing crops in the flooding zones, due to the availability of moisture and rich floodplain soils, the local people still make extensive utilisation of the floodplains for agricultural activities. Figure 3.7 shows the maize field and some corn grown on the floodplains.

Figure 3.6: Informal discussion with Chieftainess Chiawa.

Figure 3.7: Subsistence farming activities in the study area.

3.3   Physical setting

3.3.1 Geological setting

The Zambezi River basin's historical origins are said to be complex and difficult to unravel, but it is likely that the upper Zambezi was once joined the Limpopo after flowing through the Makgadikgadi pans of Northern Botswana (Axelrod and Raven 1978, Bond 1975, King 1978, Pinhey 1978, Jackson 1986, Davies 1986). The upper Zambezi is believed to have been captured by back-cutting of the Middle Zambezi, through the Batoka Gorge during the mid-Pleistocene after tectonic uplifting. This diverted the flow to the current North East direction (Beadle 1982, King 1978, Davis 1986). This theory is strongly supported by evidence from pre-impoundment fish distribution within the river (Jubb 1967, Jackson 1986), as well as other fauna elements (Pinhey 1978).

The Middle Zambezi to Cahora Bassa Dam is controlled by the west-east trending upper Zambezi graben and the middle Zambezi graben that causes the river to flow in the North West - South East direction. The stretch of the river between Kariba and Cahora Bassa lies in what is thought to be (indirectly because there is no physical connection) a branch of the Great East African Rift Valley, flanked by escarpments around 50-100 km apart. There are a series of gorges along this stretch, for example Batoka, Kariba, Mupata and at Cahora Bassa, which confine the river and greatly modify its ecology, but add to the diversity and ruggedness of its character (Davies, 1986, Timberlake, 1998,). The gorges in the Middle Zambezi basin valley are where the existing hydropower schemes have been developed and others are the potential hydropower sites such as the Batoka gorge, Devils gorge, Mupata gorge and Mepanda Nkuwa gorge (SADC/ZRA., 2008).

3.3.2 Geomorphological setting

The Middle Zambezi River reach and floodplains are situated in the geomorphological trough of the Zambezi Valley, which is bordered on both sides by escarpments of over 1,000 meters above sea level. To have a visual impression of elevation differences, a cross-section of 155 km was taken across the area from the tip of the escarpment on the Zimbabwean side to the escarpment on the Zambian side. Elevation readings were taken from the 155 km cross-section at 1km intervals from the escarpment on the Zimbabwean side to the escarpment on the Zambian side. Figure 3.8 shows the position of the cross-section and gives the cross-sectional elevation of the valley. The Zambezi escarpments border the study area. The low-lying geomorphological trough allows the formation of the middle Zambezi flood plain.

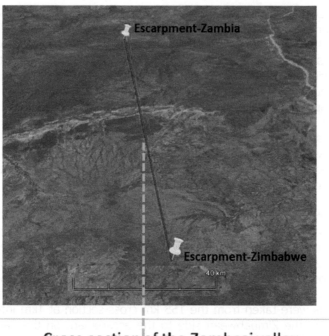

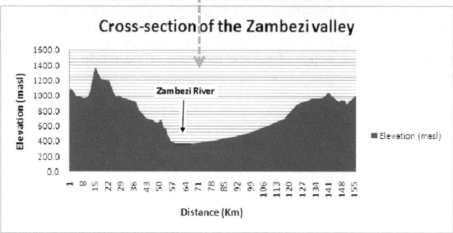

Figure 3.8: Zambezi Valley 155 km cross-section from the escarpment on Zambian side to the escarpment on the Zimbabwean (Google earth image and elevation).

3.3.3 Floodplains

The floodplain floors are gradually built up of layers of coarse materials from old streams to finer silts and clays which have dropped out of suspension onto the floodplain. Although flood plain formation is considered a long-term process, it is said that channel shifts may occur frequently enough to cause changes in habitat for aquatic and wetland species. (Gordorn et. Al, 1992). The Zambezi River has a braided appearance and comprises the following features: main river channel; secondary channels; sand bars (more temporal seasonal features); islands (more

41

permanent features); and side seasonal flooding zones (these are arranged in lower and upper terraces) and pools on the terraces (probably abandoned river channels). Refer to Figure 3.9 for the aerial view of a portion of the Middle Zambezi Floodplain.

Figure 3.9: Areal view of the Middle Zambezi Floodplain features.

The islands and the banks are typically formed of sandy alluvium overlain by a layer of clays and silts. The sands often display sedimentary structures such as cross-bedding, which show that they were deposited by the active channel. The height of this alluvium above the ordinary water level suggests that it was deposited during floods. The upper, fine-grained layer represents overbank deposits, laid on top of the floodplains and established islands during receding of floods (Figure 3.10). The different flooding zone terraces are inundated according to the size of the flood flows.

Figure 3.10: Soil stratification in the floodplain area (eroding river bank).

Since the construction of Kariba Dam, the river banks are subject to progressive erosion (Figure 3.11), which results in an important on-going channel widening process (phenomenon analyzed and quantified in Chapter 5, Section 2). Increased bank erosion can be attributed to several factors, such as river incision, which often occurs downstream of dams (Kondolf, 1997; Simon et al., 2000; Simon and Darby, 2002). Another important factor is the occurrence of fast water-level fluctuations, caused by the opening and closure of the dam gates, resulting in bank storage and bank collapse (Petts and Gurnell, 2005). Increased seepage erosion might be a result of the changed water levels in the river after the construction of Kariba Dam, resulting in increased bank failure (Wilson et al., 2007). Degradation of floodplain vegetation can result in enhanced bank erosion, considering the important protective effects that roots exert on banks (Simon and Collison, 2002; Gurnell et al., 2012).

Figure 3.11: On-going bank erosion in the study area.

3.4   Ecohydrological features

In the comprehensive assessment, based on a ranking criterion, undertaken by Timberlake in 1998, the Middle Zambezi floodplain area shows significant importance in terms of vegetation, mammals, birds, herps, fish and invertebrates (Timberlake, 1998). The floodplain is the key catalyst of the creation of the rich biodiversity that is found in the area with both the left bank and the right bank having National Game Parks and Game management areas. Table 3.1 shows the ranking of wetlands of the Zambezi Basin in which the Middle Zambezi scores the highest rank.

Table 3.1: Ranking of Wetlands of the Zambezi Basin after Timberlake, 1998.

	Vegetation	Mammals	Birds	Herps	Fish	Invertebrates	OVERALL
Barotse floodplains	3	2	2	3	3	1	2
Kafue floodplains	4	4	4	3	4	2	4
Chobe Caprivi floodplains	4	3	4	4	4	2	4
Okavango swamps	4	3	4	4	4	2	4
Lake Kariba	4	3	4	4	4	3	4
Mid-Zambezi floodplains	3	4	4	4	4	2	4
Cahora Bassa	2	2	1	2	3	1	2
Lake Malawi	3	3	4	3	4	2	3
Lower Shire wetlands	3	4	4	3	4	2	3
Zambezi Delta	1	2	2	1	1	1	1
OVERALL	2	3	4	2	3	1	

(Source: Subjective assessment based on quantity and content of available literature)
1 - insignificant/poor
2 - basic
3 - moderate
4 - good

3.4.1 Floodplain vegetation

Floodplain vegetation distribution is determined by the physical nature and age of the alluvial deposits. These can be divided into two major features: the lower and upper terraces. The riparian forests comprising the dominant species of the *Faidherbia albida* trees on both the lower and the upper terraces, with the lower terraces having younger trees and the upper terraces having the older trees (Gope, 2012). Predominant on the lower terraces are the grasses comprising the perennial grasses which include: *Vetiveria nigritana*, *Setaria spacelata*, common reeds (*phragmite*) and *Oryza barthii* while annual grasses included: *Panicum maximum*, *Echinochloa colonum* and *Urochloa trichopus* (Du Toit, 1982, Dunham, 1990a).

The upper terraces, which are the older floodplain floors have some areas of pure stands of older *Faidherbia albida* trees interspaced by areas with Natal Mahogany (*Trichelia emetica*), Sausage tree (*Kigelia africana*), *Combretum embebe*, *Figus* and brush vegetation. The predominant grasses on the upper terraces are mainly the seasonal *Echinochloa colonum* and *Urochloa trichopus* with perennial grasses such as *Vetiveria nigritana*, *Setaria spacelata*, common reeds (*phragmite)* and *Oryza barthii* found in narrow patches long the banks of the river and pools.

Recently, new vegetation species have been observed on the floodplain like *Croton megalobotrys*, and *indingofera* shrubs, which were formerly not common to the floodplains (ZPWMA, 2009). Figure 3.12 shows the different types of floodplain vegetation.

Figure 3.12: Floodplain vegetation ((A)- Lower floodplain terrace most common grasses; (B) - Reeds along river banks; (C) - Shrubs on Sandbars; (D) - Upper floodplain terrace Brush; (E) - Upper floodplain terrace grasses; (F) - Sandbar floodplain grasses; (G) - Riparian Forests)

3.4.2 Aquatic fauna

The Middle Zambezi is known to have 77 species of fish (Timberlake, 1998). Of particular importance to tourism is the Tigerfish (*Hydrocynus vittatus*) which is famous for sport fishing in the whole stretch of the river. Refer to Table 3.2 for the 77 species of fish in the Middle Zambezi. Prominent and in abundance are the Crocodiles (*Crocodylus niloticus*). Refer to Figure 3.13 for some of the aquatic fauna of the Middle Zambezi.

Figure 3.13: Some of the aquatic fauna of the Middle Zambezi (Tigerfish, Redbreasted Tilapia and Crocodile).

Table 3.2: The Fish species of the Middle Zambezi (Skelton, 1994 and Timberlake, 1998).

NAME OF FISH	NAME OF FISH
Protopterus amphibius	Brycinus imberi
Protopterus annectens	Brycinus lateralis (Striped robber)
Hippopotamyrus ansorgii	Hemigrammopetersius barnardi
Hippopotamyrus discorhynchus	Hydrocynus vittatus (Tigerfish)
Marcusenius macrolepidotus	Micralestes acutidens
Mormyrops anguilloides	Schilbe intermedius
Mormyrops longirostris	Amphilius natalensis
Petrocephalus catostoma	Amphilius uranoscopus
Kneria auriculata	Leptoglanis rotundiceps
Barbus afrohamiltoni	Clarias gariepinus
Barbus annectens	Clarias ngamensis
Barbus atkinsoni	Clarias theodorae
Barbus choloensis	Heterobranchus longifilis
Barbus eutaenia	Malapterurus electicus
Barbus fasciolatus	Chiloglanis emarginatus
Barbus haasianus	Chiloglanis neumanni
Barbus kerstenii	Chiloglanis pretoriae
Barbus lineomaculatus	Synodontis nebulosus
Barbus macrotaenia	Synodontis zambezensis
Barbus manicensis	Aplocheilichthys johnstoni
Barbus marequensis	Aplocheilichthys hutereaui
Barbus miolepis	Aplocheilichthys katangae
Barbus multilineatus	Nothobranchius orthonotus
Barbus paludinosus	Nothobranchius rachovii
Barbus radiatus	Aethiomastacembelus shiranus
Barbus toppini	Astaotilapia calliptera
Barbus trimaculatus	Oreochromis macrochir (longfirn tilapia)
Barbus unitaeniatus	Oreochromis mortimeri (Kariba tilapia)
Barbus viviparus	Oreochromis mossambicus
Labeo altivelis	Oreochromis placidus
Labeo congoro	Oreochromis shiranus
Labeo cylindricus	Pharyngochromis acuticeps
Labeo molybdinus	Pseudocrenilabrus philander
Opsaridium zambezense	Sargochromis codringtonii
Opsaridium sp.	Tilapia rendalli (Redbreast tilapia)
Varicorhinus nastus	Tilapia sparrmanii (Banded tilapia)
Varicorhinus pungweensis	Ctenopoma intermedium
Distichodus mossambicus	Ctenopoma multispine
Distichodus schenga	

3.4.3 Terrestrial fauna

The Middle Zambezi floodplains provide a rich habitat that supports abundant populations of wildlife. Table 3.3 outlines the large mammals of the study area. Figure 3.14 shows some of the commonly seen animals of the Middle Zambezi.

Table 3.3: Large Mammals of the Middle Zambezi (Jarman, 1972).

Common Name	Scientific Name
Vervet monkey	*Cercopithecus aethiops L.*
Baboon	*Papio ursinus Kerr*
Jackal	*Canis adustus Sundevall*
Wild dog	*Lycaon pictus Temminck*
Spotted hyaena	*Crocuta crocuta Erxleben*
Leopard	*Panthera pardus L.*
Lion	*P. leo L.*
Elephant	*Loxodonta africana Blumenbach*
Black rhinoceros	*Diceros bicornis L.*
Zebra	*Equus burchelli Gray*
Bush pig	*Potamochoerus aethiopicus Pallas*
Hippopotamus	*Hippopotamus amphibius L.*
Common duiker	*Sylvicapra grimmia L.*
Grysbok	*Raphicerus Sharpei Thomas*
Klipspringer	*Oreotragus oreotragus Zimmermann*
Reedbuck	*Redunca arundinum Boddaert*
Waterbuck	*Kobus ellipsiprymnus Ogilby*
Impala	*Aepyceros melampus Lichtenstein*
Roan antelope	*Hippotragus equinus Desmarest*
Sable antelope	*Hippotragus niger Harris*
Bushbuck	*T. scriptus Pallas*
Nyala	*T. angasi Gray*
Greater kudu	*Strepsiceros strepsiceros Pallas*
Eland	*Taurotragus oryx Pallas*
Buffalo	*Syncerus caffer Sparrman*

Figure 3.14: Common large mammals seen in the Study area.

3.4.4 Threats

Despite the fact that the Study area is well acknowledged for its beauty and rich biodiversity, the Middle Zambezi does face real threats towards its sustainability. Timberlake (2000) acknowledges that the construction of dams has had probably the greatest effect on biodiversity of wetland and aquatic species and on wetland ecological processes. The main threats lay in the changes in hydrology, modification of flooding regimes, and hence lead to negative consequences to the habitat and species composition.

3.4.4.1   Poaching

The threat of poaching may lead to depletion of wildlife in the area. To counter this threat of poaching, the area on both the Zambian and Zimbabwean side does have well established systems for conservation of wildlife. The National Parks Authorities have the mandate to conserve the habitat and wildlife. The Conservation efforts by the National Agencies are being supplemented by Community based initiatives, noteworthy is the Conservation Lower Zambezi (CLZ), which has scored a lot of accolades in supplementing the efforts of the Zambia Wildlife Authority.

3.4.4.2   Transformation of floodplain wetlands in drier terrestrial environment
As the floodplain is located downstream of the Middle Zambezi hydropower schemes, most of the flows through this river reach are regulated depending on the hydropower schemes needs. This threat of floodplain transformation from wetland to terrestrial environment is well acknowledged by the researchers and conservation agencies (Attwell, 1970, Dunham, 1989). This arises from the infrequent or a complete lack of flooding due to water regulation.

This threat can only be resolved by the National water resource Managers and Policy Makers working closely with the dam operators to come up with ways of crafting flow regimes that would allow for deliberate occasional flooding.

3.4.4.3   Increased pressure of overgrazing
The floodplain ecosystems are sustained by the seasonal circle of flooding and recession. The flooding of some portions of the floodplains enables animal migration to the drier parts of the floodplain, giving relief to the flooded portions. As flood recession occurs after a number of months, the animals are able to migrate back to the areas that were flooded. This circle is important to allow for growth and regeneration of floodplain vegetation. In the absence of flooding however, the animals would be resident on the floodplains for all the months of the year, this puts undue pressure on the edible vegetation leading to overgrazing and lack of vegetation regeneration.

CHAPTER 4: SURFACE AND SUBSURFACE FLOWS

This chapter describes the results of field investigations (surface and sub-surface water level time series, cross-sectional and longitudinal bed level variations and cross-sectional flow velocity distribution at some specific locations) and hydrodynamic modeling to assess the hydraulic behavior of the Middle Zambezi River system. The work starts with a hydrological analysis, based on the water balance, including the assessment of a correlation between dam operations and observed water level fluctuations. The availability of new data acquired during the study allowed setting up a 1D hydrodynamic model to basically establish a relation between discharge and extent of inundated area (wetted width) and a simple sub-surface flow model to establish the direction of sub-surface flow (towards the river or towards the floodplain). The work in this chapter has benefited from the MSc research by Khan (2013) and part of it has been published in Khan, Mwelwa-Mutekenya et al. (2014).

4.1   Hydrological set-up

Although the study area is located downstream of the major Zambezi basin hydropower schemes, there is absence of an existing hydrometric network to facilitate measurements and monitoring. Therefore, for this research, it became imperative to carry out field observations to facilitate understanding of the hydrological forcing and response.

4.1.1 Field measurements: water levels and temperature
A network of 8 hydrometric stations was established during the research period. All the stations were equipped with staff gauge plates and divers were also installed at five of the stations. Table 4.1 outlines the established hydrometric network and the equipment installed at each station. The location and lay-out of the hydrometric network is as shown in Figure 4.1. Monitoring of daily water levels involved installation of a combination of divers and manual staff gauge plates. The gauge plates were read three times a day (06:00 hours, 12:00 hours and 18:00 hrs) while divers were set at 5 minutes recording interval, recording both water level and temperature variation (refer to Figure 4.2, showing the water level monitoring equipment and Figure 4.3, showing the diver data spreadsheet). For monitoring of the subsurface water flow, a well was constructed and installed with a diver. Figure 4.4 shows the subsurface water flow monitoring process.

Table 4.1: Hydrometric stations in the study area (locations are mapped in Figure 4.1)

No.	Name of Station	Latitude (S)	Longitude (E)	Elevation (m asl)	Monitoring Equipment
ZM1	Zambezi River at Namoomba	16°21′01.1″	028°50′13.0″	386	Gauge Plates
ZM2	Zambezi River at Chirundu Bridge	16°02′16.3″	02851′02.8″	373	Q measurement only-
KF	Kafue River at Gwabi	15°57′05.5″	028°51′38.9″	380	Gauge Plates / Diver
ZM3	Zambezi River at Chakanaka Farm	15°56′20.1″	028°57′03.1″	367	Gauge Plates / diver / baro diver
ZM4	Zambezi River at Samango Camp	15°45′53.1″	029°13′20.1″	367	Gauge Plates / Diver
ZM5	Zambezi at Chiawa Camp	15°41′05.8″	029°24′47.9″	357	Gauge Plates / Diver / well Diver
MP1	Mana River at Long Pool	15°44′51.5″	029°21′20.1″	359	Diver
MP2	Mana River at Mana bridge	15°44′20.7″	029°22′04.8″	359	Diver
LK	Lukumechi river at bridge	16°03′28″	29°24′28″	473	sediment
Q1	Kariba Dam	16°31′18″	28°45′41″	488	Inflow1(main)
Q2	Kafue Gorge Dam	15°48′30″	28°25′15″	994	Inflow2(tributary)
Qout	Mupata Gorge	15°38′13″	30°01′53″	337	Out flow

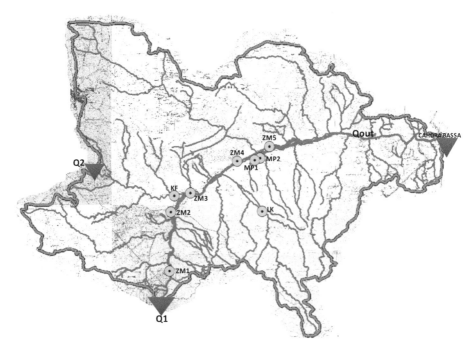

Figure 4.1: Hydrometric network in the study area.

Figure 4.2: Surface water level measurement and monitoring.

Identification	=TEMPERATURE
Reference level	=-20.00 °C
Range	=100.00 °C

[Data]

CHAKANAKA FARM BARO DIVER DATA			GWABI LODGE DIVER DATA				CHAKANAKA FARM DIVER DATA				CHIAWA CAM
Date/time	Pressure[n	Temperature[°C]	Date/Time	Pressure	Temp	WL(m)	Date/Time	Pressure	Temp	WL(m)	Date/Time
28/11/2010 11:00	9.843	34.99					28/11/2010 11:00	10.284	26.96	0.441	28/11/20
28/11/2010 11:05	9.844	34.99					28/11/2010 11:05	10.28	26.97	0.436	28/11/20
28/11/2010 11:10	9.841	35.22					28/11/2010 11:10	10.285	26.99	0.444	28/11/20
28/11/2010 11:15	9.839	35.42					28/11/2010 11:15	10.287	27.01	0.448	28/11/20
28/11/2010 11:20	9.839	35.5					28/11/2010 11:20	10.277	27.03	0.438	28/11/20
28/11/2010 11:25	9.837	35.62					28/11/2010 11:25	10.283	27.05	0.446	28/11/20
28/11/2010 11:30	9.837	35.74					28/11/2010 11:30	10.275	27.07	0.438	28/11/20
28/11/2010 11:35	9.835	35.93					28/11/2010 11:35	10.277	27.09	0.442	28/11/20
28/11/2010 11:40	9.835	36.07					28/11/2010 11:40	10.27	27.11	0.435	28/11/20
28/11/2010 11:45	9.834	36.29					28/11/2010 11:45	10.262	27.13	0.428	28/11/20
28/11/2010 11:50	9.834	36.42					28/11/2010 11:50	10.263	27.14	0.429	28/11/20
28/11/2010 11:55	9.834	36.5					28/11/2010 11:55	10.265	27.16	0.431	28/11/20
28/11/2010 12:00	9.834	36.63					28/11/2010 12:00	10.26	27.18	0.426	28/11/20
28/11/2010 12:05	9.83	36.82					28/11/2010 12:05	10.263	27.19	0.433	28/11/20
28/11/2010 12:10	9.832	36.87					28/11/2010 12:10	10.263	27.2	0.431	28/11/20
28/11/2010 12:15	9.83	37.04					28/11/2010 12:15	10.267	27.21	0.437	28/11/20
28/11/2010 12:20	9.83	37.2					28/11/2010 12:20	10.255	27.23	0.425	28/11/20
28/11/2010 12:25	9.83	37.38					28/11/2010 12:25	10.257	27.25	0.427	28/11/20
28/11/2010 12:30	9.828	37.47					28/11/2010 12:30	10.254	27.27	0.426	28/11/20

Figure 4.3: Diver data spreadsheet showing data for baro diver and diver for Zambezi River at Chakanaka (ZM3).

Figure 4.4: Subsurface flow measurement and monitoring at ZM5.

4.1.1.1   Diurnal water level variation

Most of the flows in the river reach are turbine outflows from Q1(Kariba) and Q2(Kafue Gorge), as outlined in Table 4.1 and Figure 4.1. The diver water levels at readings of 5-minute intervals revealed a clear trend of diurnal water level variation (within a range of 15 cm) related to the power generation needs with higher levels in the morning, a lower trough in the afternoon and a steady rise in the night. The water levels significantly vary on the 24-hour circle, as can be seen in the levels for the hydrometric station of Zambezi at Chakanaka Farm (ZM3), as shown in Figure 4.5. The same trend is confirmed with the manual gauge plate readings (gauge readers read the gauge plate at 06:00 hours, 12:00 hours and 18:00 hours), with the readings as shown in Figure 4.6.

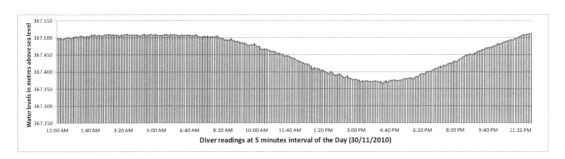

Figure 4.5: Diurnal water level fluctuation according to generation needs at ZM3 hydrometric station.

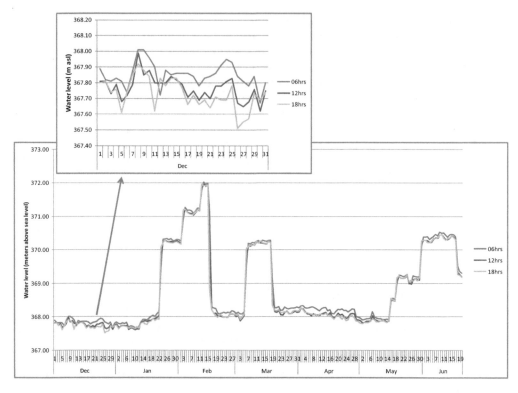

Figure 4.6: Gauge plate readings showing the diurnal water fluctuation at ZM3 (readings from December 2010 to June 2011).

4.1.1.2   Flood spillage

Though the Kariba spillway gates are rarely opened, the research period coincided with the extremely wet rainy period that led to the filling of the Kariba, bringing the necessity of spilling. During the period of spillage an extreme range of 0 to 5 m of water level fluctuation was observed, as the opening and closing of flood gates were not phased. This is an extreme departure from the natural system and may lead to negative consequences in terms of ecology and morphology (Brown and King, 2011, Khan et al. 2014). A detailed analysis was carried out over the period January to March 2011, when the flood gates were suddenly opened preceded by a long time without flood spillage from Kariba. Figure 4.7 shows a step wise hydrograph depicting the following gate openings: two gates (4,000 m^3/s), then three gates (5,500 m^3/s) and four gates (7,000 m^3/s). All the gates were closed at once on February 16th 2011, leading to a drop in discharge to about 1,000 m^3/s (Figure 4.7), translating to about 5 metres drop in water levels within a period of 24 hours, as monitored at ZM3. Figure 4.8 shows the response of the Middle Zambezi River reach in terms of water level change (about 4 metres drop) to the impromptu closing of the four gates.

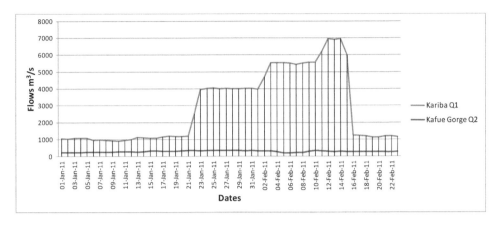

Figure 4.7: Outflows from Kariba (Q1) and Kafue Gorge Q2).

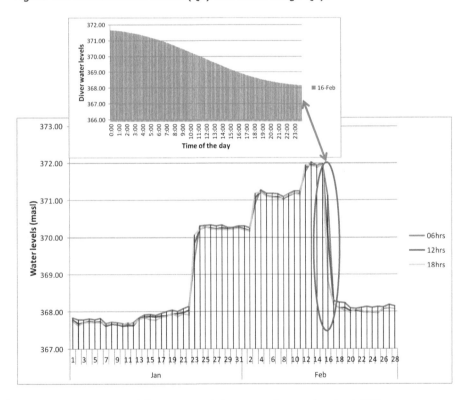

Figure 4.8: Response of the river reach in terms of water levels at ZM3.

The quick closure of spillway gates, leading to a drop in water levels on the order of 5 m in one day, has likely caused important destabilizing effects on the river banks. This is supported by the findings of previous research (e.g. Wolman, 1959; Hooke, 1979; Thorne, 1982). Rinaldi et al. (2004) show that the probability of bank failure increases by the presence of more than one discharge peaks in a relatively short time interval. Figure 4.9 shows that this occurred in the years 2000-2001 and 2010-2012.

4.1.1.3 Long-term hydrological variations

The long-term hydrological variations show a ten-year cycle of wet and dry periods. The analysis of the flow record from 1905 to 2012, and superimposing the hydropower flow regulation, shows that three distinct hydrological periods emerge. The three periods are as follows:

a) Pre-Kariba (1905 to 1959): unregulated natural system characterised by annual peaks in the rainy season and low flows in the dry season

b) Post-Kariba 1 (1959 to 1978): regulated flows with only Kariba North Bank Power Station operating, the period coinciding with a wet cycle and is characterised by annual dam spill flows

c) Post-Kariba 2 (1979 to date): regulated flows mainly turbine flows from three power stations (Kariba North Bank, Kariba North Bank and Kafue Gorge). This period typically presents extensive time intervals of no dam spill flows as the period coincides with a dry cycle. However, at the same time this period is characterised by higher regular low flows. This indicates that even in the natural conditions (absence of dams) there were dry years without sufficient flooding with some exceptionally wet periods.

After the implementation of the Kariba and (later) Kafue hydropower schemes, the water regulation variation results in the three hydrodynamic periods: pre-Kariba; post-Kariba 1 and post-Kariba 2. Inflows into the Kariba, as monitored from Zambezi River at Victoria falls, represent the unregulated flows, while the flows from Kariba represent the regulated flows. Figure 4.9 shows the three distinct hydrological periods of the Middle Zambezi.

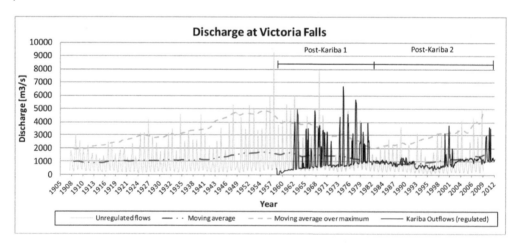

Figure 4.9: Three distinct hydrological periods identified in the River reach from 1905 to 2012 (Khan et al., 2014).

The difference in the hydrological state of the three periods is also shown in the Figure 4.10 (flow-duration curves) and 4.11 (mean annual discharge hydrographs).

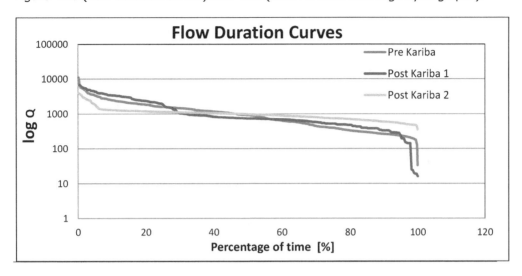

Figure 4.10: Flow duration curves for the three distinct hydrological periods of Middle Zambezi.

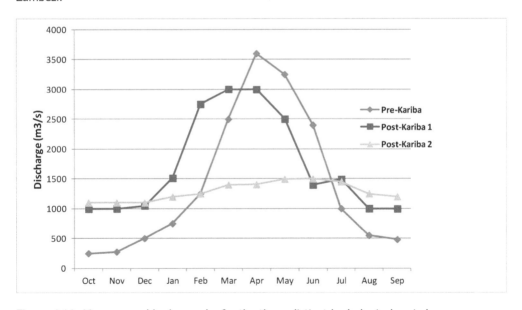

Figure 4.11: Mean annual hydrographs for the three distinct hydrological periods.

4.1.2 Field measurements: longitudinal and cross-sectional flow velocity

An important part of the field research campaign was the measurement of the longitudinal and flow velocity cross-sections of the Middle Zambezi River. The

Acoustic Doppler Current Profilers (ADCP) equipment which was mounted on a speed boat with a team of four people was used, as can be seen in Figure 4.12.

4.1.2.1   Discharge measurement

The discharge measurement were carried out for low flows (November 2010) and high flows (March 2011) for ZM1, ZM2, ZM3 and ZM4 and KF. The boat runs were done four times to get the best cross-sectional measurement. The total discharge through a measurement section was computed based on the mean water velocity in the water column and the cross-sectional area. For the purposes of measurement, the cross section was broken into three key components: the start edge, the transect and the end edge (Figure 4.13). The area and velocity of these components were then summed to calculate the total discharge. Although the ADCP have been used since the late 1980's to measure discharge in rivers and comprehensive studies of this technique have shown that the use of ADCPs from moving vessels produces reliable discharge measurements under most circumstances tested according to SonTek (2010) and Gordon (1989), when the spreadsheets of the measurement were checked, it was found that for all the measurements, the program was underestimating the discharge measured by excluding the areas near the banks (Figure 4.13: Start Edge and End Edge). The near-bank discharge was then calculated and adjusted for all measurements to give the correct measurements, as shown in Figure 4.14. Apart from flow discharge, the SonTek program gives a number of important cross sectional parameters, including the graphical cross sectional velocity (Figure 4.15).

Figure 4.12: ADCP field discharge measurement and longitudinal profile measurement process.

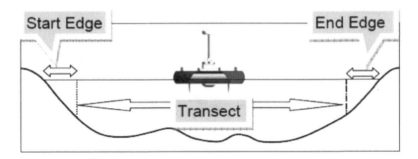

Figure 4.13: Graphical presentation how the cross-section was divided when carrying out the discharge measurement (start edge, transect and end edge).

ZM4 Q measurement on 27th November 2010 at 14.24

Step	Sample	Time	Track m	Length m	DMG m	Depth m	# Pings	# Cells	Satellites	GPS Qualit	Mean Spee m/s	Boat Spee m/s	Left Q m3/s	Right Q m3/s	Total Q m3/s
Start Edge	1	5:31:45 AM	1.34	1.375	1.34	1.59	31	11	0	0	0.711	1.342	0	0	1.554
Start Edge	2	5:31:46 AM	2.75	1.385	2.75	1.66	31	12	0	0	0.673	1.406	0	0	1.547
Start Edge	3	5:31:47 AM	4.11	1.37	4.1	1.59	32	11	0	0	0.909	1.361	0	0	1.980
Start Edge	4	5:31:48 AM	5.49	1.395	5.48	1.62	31	11	0	0	0.788	1.38	0	0	1.781
Start Edge	5	5:31:49 AM	6.9	1.3	6.9	1.7	32	12	0	0	0.936	1.416	0	0	2.069
Start Edge	6	5:31:50 AM	8.09	1.205	8.07	1.54	32	11	0	0	0.607	1.183	0	0	1.126
Start Edge	7	5:31:51 AM	9.31	1.12	9.3	1.56	31	11	0	0	0.709	1.227	0	0	1.401
Start Edge	8	5:31:52 AM	10.33	0.945	10.27	1.65	32	12	0	0	0.709	1.015	0	0	1.106
Start Edge	9	5:31:53 AM	11.2	0.77	11.07	1.71	31	12	0	0	0.631	0.875	0	0	0.831
Start Edge	10	5:31:54 AM	11.87	0.625	11.54	1.71	32	12	0	0	0.97	0.667	0	0	1.037
Start Edge	11	5:31:55 AM	12.45	0.545	11.6	1.79	31	13	0	0	0.727	0.58	0	0	0.709
Start Edge	12	5:31:56 AM	12.96	0.47	11.36	1.74	31	12	0	0	0.935	0.511	0	0	0.765
Start Edge	13	5:31:57 AM	13.39	0.525	10.98	1.8	32	13	0	0	0.891	0.425	0	0	0.842
Start Edge	14	5:31:58 AM	14.01	0.6	10.36	1.74	31	12	0	0	0.852	0.625	0	0	0.889
Start Edge	15	5:31:59 AM	14.59	0.635	9.8	1.79	32	12	0	0	0.806	0.573	0	0	0.916
Start Edge	16	5:32:00 AM	15.28	0.745	9.2	1.7	31	12	0	0	0.895	0.691	0	0	1.134
Start Edge	17	5:32:01 AM	16.08	0.825	8.58	1.72	32	12	0	0	0.894	0.803	0	0	1.269
In Transec	18	5:32:02 AM	0.85		0.85	1.74	31	12	0	0	0.874	0.847	0	0.99	2.56	20.955
In Transec	19	5:32:03 AM	1.82		1.81	1.6	32	11	0	0	0.827	0.974	0	1.27	3.61
In Transec	393	5:38:17 AM	702.55		585.08	2.25	31	17	0	0	0.465	1.336	0	1.98	1,235.87
End Edge (	394	5:38:18 AM	703.85		585.53	2.06	31	14	0	0	0.449	1.297	1.2	1.99	1,237.08
End Edge (	395	5:38:19 AM	704.97		585.97	1.77	31	13	0	0	0.399	1.128	1.31	1.99	1,237.19
End Edge (	396	5:38:20 AM	706.05	1.02	586.38	1.85	31	13	0	0	0.293	1.075	1.27	1.99	1,237.15	0.553
End Edge (	397	5:38:21 AM	707.01	0.945	586.7	1.95	31	14	0	0	0.249	0.956	1.19	2	1,237.07	0.459
End Edge (	398	5:38:22 AM	707.94	0.9	587.06	2	31	15	0	0	0.462	0.93	1.27	2	1,237.15	0.832
End Edge (	399	5:38:23 AM	708.81	0.83	587.35	2.03	31	15	0	0	0.314	0.871	1.27	2	1,237.16	0.529
End Edge (	400	5:38:24 AM	709.6	0.805	587.6	2.04	31	15	0	0	0.37	0.79	1.28	2	1,237.18	0.608
End Edge (	401	5:38:25 AM	710.42	0.805	587.71	2.03	31	15	0	0	0.289	0.821	1.28	2.01	1,237.17	0.472
End Edge (	402	5:38:26 AM	711.21	0.805	587.55	1.99	31	15	0	0	0.195	0.788	1.22	2.01	1,237.11	0.312
End Edge (	403	5:38:27 AM	712.03	0.765	587.21	1.97	31	14	0	0	0.25	0.822	1.16	2.01	1,237.06	0.377
End Edge (	404	5:38:28 AM	712.74	0.705	586.72	1.94	31	14	0	0	0.343	0.71	1.15	2.01	1,237.06	0.469
End Edge (	405	5:38:29 AM	713.44	0.685	586.13	1.94	31	14	0	0	0.236	0.671	1.08	2.01	1,236.99	0.314
End Edge (	406	5:38:30 AM	714.11	0.665	585.5	1.93	31	14	0	0	0.255	0.671	1.05	2.01	1,236.95	0.327
End Edge (	407	5:38:31 AM	714.77	0.685	584.87	1.97	31	14	0	0	0.225	0.667	0.99	2.01	1,236.89	0.304
End Edge (	408	5:38:32 AM	715.48	0.68	584.16	1.99	31	15	0	0	0.315	0.71	0.97	2.01	1,236.87	0.426
End Edge (	409	5:38:33 AM	716.13	0.325	583.52	1.99	31	15	0	0	0.263	0.649	0.91	2.01	1,236.82	0.170
															6.151	1264.30

Figure 4.14: ADCP spreadsheet underestimates the discharge by excluding the areas of start edge and end edge. The last column shows the manual correction - $Q_{(uncorrected)}$ = 1,236.8 m³/s and $Q_{(corrected)}$ = 1,264.3 m³/s.

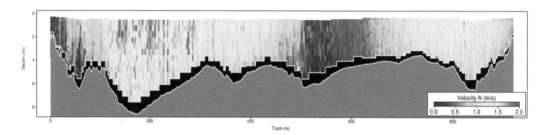

Figure 4.15: ZM1 cross sectional flow velocity (high flows) measured on 9th March 2011 at 11:00 hours with Q= 4,067.63 m³/s.

4.1.2.2 Longitudinal profile for the Middle Zambezi

A total of 79,965 km of longitudinal profile of the Middle Zambezi was measured following the navigable parts of the river channel. According to the specifications of ADCP (SonTeck, 2010) the river profile is to be measured from downstream going in the upstream direction. Therefore, the profile measurement started from ZM5 to ZM1 with the boat moving against the current. However, in the presentation of the results, the profiles have to be flipped so that the segments are from upstream to downstream (ZM1 to ZM5), Table 4.2 Shows the river reach segments in the order of measurement and in the order of presentation, with the distances covered. Between ZM1 and ZM2, there are some areas which were too shallow and therefore the complete segment from ZM2 to ZM1 could not be measured in its entirety, only a

7 km segment was measured from navigable area to the Kariba Gorge. The flipped segment 1 (from Kariba Gorge, past ZM1 to the shallow points upstream of ZM2) is shown in Figure 4.16. The profile shows significant river morphological features where the segment is deepest, the river presents a single thread channel and where the segment is shallow it presents a higher degree of breading.

Table 4.2: Middle Zambezi longitudinal profiles measured going in the upstream direction from MZ5 towards ZM1- the boat moving against the water current (however in presentation, the profile has to be flipped).

Measurement Segment No.	From	To	Distance (Km)	Presentation flipped segment No.
1	ZM5	ZM4	23,896.11	5
2	ZM4	Mid point (upstream of ZM4)	11,235.31	4
3	Mid point - upstream of ZM4	ZM3	27,246.89	3
4	ZM3	ZM2	10,586.55	2
5	Shallow point (upstream of ZM2)	Kariba Gorge (upstream of ZM1)	7,000.00	1
Total distance covered			79,964.86	

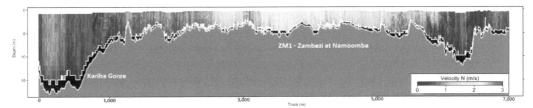

Figure 4.16: Segment 1 of flipped river reach of the Middle Zambezi longitudinal profile from Kariba Gorge (7 km profile from Kariba Gorge through ZM1 to the shallow areas upstream of ZM2).

4.1.3 Surface water balance of the study area

4.1.3.1   Water balance concept

The water balance aids the understanding of inflow, outflow and storage in the study sub-catchment of the Zambezi River. The approach is widely used in water resources management strategies in order to account for water availability, variability and consumption. Well undertaken water balance studies are able to bring to the fore vital statistics for use in water accounting and decision making process (Soulard, 2003; United States Department of the Interior, 2000). Based on water balances, it is possible to make quantitative evaluations of water resources and their changes under the influence of human activities. The knowledge of water balances assists in the prediction of the consequences of changes to the hydrological regime of the basin. In this study, the water balance also assists in the hydrological simulations, providing the foundation for checking the validity of the data generated from the models (Sokolov and Chapman, 1974).

The Water Balance is defined as the systematic presentation of data on the supply and use of water within a geographical region for a specific period. The water balance of a basin assumes that all water entering the basin during a given period of time must either go into storage within the boundaries, be consumed or leave the domain (National Institute of Hydrology, 1999). In general, the water balance equation simplifies a very complex hydrological cycle for a particular area to a short expression where runoff (Q) plus or minus change in storage ΔS is equal to precipitation (P) minus evaporation (E) from land and water surfaces. A general water balance equation can be presented as follows:

$$Q + \Delta S = P - E \tag{4.1}$$

In which

Q = discharge [m^3/s]
ΔS = losses due to storage [m^3/s]
P = precipitation from local catchment [m^3/s]
E = losses due to evaporation [m^3/s]

With water balance data, it is possible to compare individual sources of water in a system over different periods of time and establish their effects on variations in water regime. The analysis used to compute individual water balance equation components and the coordination of these components in the water balance equation can make it possible to identify deficiencies in the distribution of observation stations and to discover systematic errors in measurements. The water balance also provides an indirect evaluation of an unknown component from the difference between the known components. In the case of this study, the contribution of the local catchment tributaries, which is unknown, is estimated and the estimate used for the calibration of rainfall runoff models that are built for the ungauged tributaries.

Sokolov and Chapman (1974) indicate that all water balance components are subject to errors of measurement or estimation and propose that each water balance equation should include a discrepancy term (μ). However in our computation for the Middle Zambezi we would consider this to be negligible as most of the components are measured components and the size of the sub-catchment is small.

4.1.3.2 Assessment of the water balance

For the hydropower dominated sub-catchment of the Middle Zambezi basin, the following equation has been adopted from the general water balance equation after Gupta, 1989:

$$P + Q_{I(1)} + Q_{I(2)} - E - Q_O - \Delta S - Q_{irr} = 0 \tag{4.2}$$

Where

P is the total input due to precipitation from the catchment - this is here treated as the contribution of the local catchment tributaries [m³/s].

$Q_{I(1)}$ is the discharge outflow from Kariba reservoir, treated as discharge inflow (1) into the system [m³/s].

$Q_{I(2)}$ is the discharge outflow from Kafue Gorge reservoir, treated as inflow (2) into the system [m³/s]

E is total loss due to evaporation from the river channel water surface [m³/s].

ΔS is the total loss due to storage within the river reach and flood plain [m³/s].

Q_O is the discharge outflow at the last monitoring station [m³/s].

Q_{irr} is the total water abstraction for irrigation [m³/s].

The water balance system and how the components are computed are outlined in Table 4.3.

Table 4.3: Water Balance Components and mode of calculation.

Symbol	Specification	Source of data	Measured	Calculated / modeled
P	Runoff contribution of the local catchment tributaries			√
$Q_{I(1)}$	Discharge inflow (1) into the system from Kariba	Zambezi River Authority		
$Q_{I(2)}$	inflow (2) into the system from Kafue Gorge	ZESCO		
E	river channel water surface			√
ΔS	storage within the river reach and flood plain			√
Q_{irr}	Water abstraction for irrigation			√
Q_O	Discharge outflow at the last monitoring station		Nov. 2010 March 2011	

4.1.3.3   Limitations on the comparison of discharge measurement and the daily averaged reservoir outflows.

The discharge measurements are instantaneous flows at the particular time of the measurement whereas the outflows from the reservoirs are a daily average. The daily average does not show the diurnal variation due to power generation fluctuation depending on power demand at that particular time. The diurnal fluctuation of flows due to the power generation fluctuation can be seen clearly from Figure 4.17, showing the water level data recorded by the divers every five minutes. The discharge measurement at (ZM3) was made at 03:18 PM at a time when power generation is off peak and hence the reduced generation that corresponds with reduced turbine discharge (Figure 4.17).

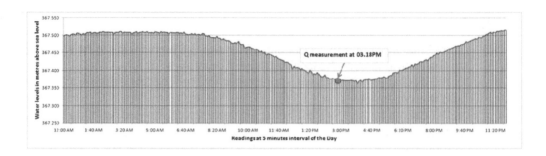

Figure 4.17: ZM3 diurnal variation of water levels corresponding to power generation demand (the discharge measurements were done at 03:18PM which is in the afternoon lower trough of power generation).

4.1.3.4   Limitation of the quality of the discharge measurement

The discharge measuring process is always an estimation of reality. One of the aspects of the measuring process is that the areas close to the banks are always estimated and no actual measurement is done. The competence of the field team and the accuracy of the equipment used during the process is a key aspect to determine how close one gets to reality for each measurement. On the other hand the reservoir releases are not an estimate but well known as the spill gates are well calibrated and the water usage of each turbine is well known and monitored during the power generation process.

Upon close examination of all the discharge measurement spreadsheets, it was found that the equipment did not add the discharge in the areas close to both banks, even when the lengths, depths and velocity of the specific areas were measured. In some cases the equipment was subtracting from the discharge instead of adding. In all these cases the necessary recalculations were done and in some cases the differences were quite significant as can be seen in Table 4.4

Upon adjustment of all the figures, it was decided that the adjusted figures would be used in this study. The difference between the ADCP discharge readings and the adjusted figures range from 16 m^3/s to 291 m^3/s. With such a wide range, it is important that each time a measurement is made, the spreadsheet should be checked to verify the discharge result. The results from the ADCP should be critically examined to correct any systematic errors, (Figure 4.14).

Table 4.4: Corrected ADCP Discharge measurement figures.

No.	Station	Date and Time	ADCP Program Q calculation	Corrected Q figures	Difference
1	ZM1	30/11/10 - 13:05 hours	1,213.10	1,384.29	171.19
2	ZM1	09/03/11 - 11:44 hours	4,027.98	4,067.63	39.65
3	ZM2	29/11/10 - 14:05 hours	1,083.81	1,099.99	16.18
4	ZM2	09/03/11 - 16:57 hours	3,841.33	4,132.29	290.96
5	ZM3	28/11/10 - 15:18 hours	1,256.85	1,275.03	18.18
6	ZM3	09/03/11 - 18:12 hours	4,185.63	4,213.02	27.39
7	ZM4	27/11/10 - 14:24 hours	1,236.82	1,264.30	27.48
8	ZM4	10/03/11 - 13:23 hours	4,273.16	4,364.35	91.19

4.1.3.5 Comparison of discharge measurements with the daily average reservoir outflows for stations upstream of the Kafue River confluence.

The discharge at ZM1 (Table 4.4) of 1,384 m^3/s was measured on November 30th 2010 at 13:05. Its value is higher than the daily average Kariba outflow of 1,111 m^3/s because the measurement time coincided with the midday generation peak. The discharge measurement at ZM2 of 1,100 m^3/s was done on November 29th 2010 at 14:05, compares very well with the daily average Kariba outflow of 1,090 m^3/s.

For flood monitoring, the discharge measurements at ZM1 and ZM2 were carried out on March 9th 2011. They all compare very well to the average daily reservoir outflow of 4,086 m^3/s. The discharge measurement at ZM1 was 4,068 m^3/s, while the discharge measurement at ZM2 was 4,132 m^3/s. The disparity of 18 m^3/s, between the average Q1 outflow and measured discharge at ZM1 and the difference in discharge of 46 m^3/s measured at ZM2 compare very well. It can be pointed out that station ZM1 is just 27 km downstream of Kariba, where most of the river channel is through the Kariba gorges with no flood plains. The measurement was carried out at 11:44 within the morning peak generation. On the other hand, the discharge measurement at ZM2 was carried out at 04:39 PM within the afternoon trough (lower generation) and the station is 66 km from Kariba. Figure 4.18 shows the November 2010 and March 2011 hydrographs and the discharge measurements.

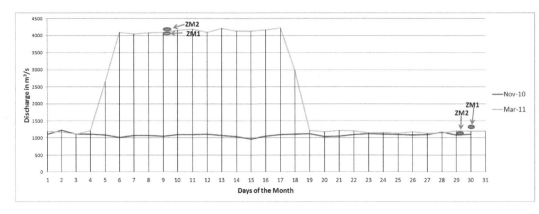

Figure 4.18: Flows and discharge measurements made upstream of Kafue River confluence.

4.1.3.6 Comparison of discharge measurement with the daily average reservoir outflows for stations downstream of the Kafue River confluence

The stations downstream of the confluence of the Kafue River benefit from the outflows from Kafue Gorge reservoir (Q2). Therefore, the flows downstream of confluence are expected to be higher due to the contribution of Kafue River.

The discharge at ZM3 on 28th November 2010 at 15:18 was 1,275 m^3/s against a combined Q1 and Q2 discharge of 1,394 m^3/s, representing losses of 119 m^3/s. The discharge measured at ZM4 on 27th November 2010 at 14:16 was 1,264 m^3/s, against a combined Q1 and Q2 discharge of 1,303 m^3/s, representing a loss of 39 m^3/s. Figure 4.19 shows the November 2010 hydrograph and the discharge measurements.

The discharge at ZM3 on 9th March 2011 at 18:12 resulted equal to 4,213 m^3/s, against a combined Q1 and Q2 discharge of 4,454 m^3/s, resulting in losses of 241 m^3/s. Losses of this magnitude are justifiable in a river reach which is braided with the width of the channel extending for over 2 kilometers in some sections. Flooding of the flood plains contributes to an increase in storage.

The discharge measured at ZM4 on 10th March 2011 at 13:23 was 4,364 m^3/s, against a combined Q1 and Q2 discharge of 4,519, resulting in a loss of 155 m^3/s. This loss can be attributed to the extensive flooding in the river reach upstream of ZM4, as the channel section at ZM4 is the only single section in this area. More downstream, as one approaches Mana Pools and Chiawa camp, the channel is highly braided with multiple channels which make it difficult to measure the discharge. Figure 4.19 shows the March 2011 hydrograph and the discharge at ZM4.

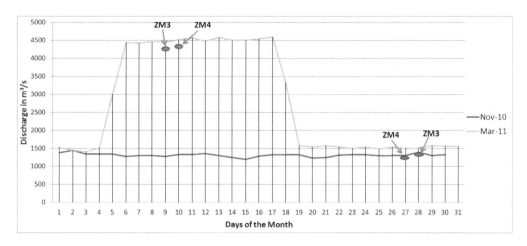

Figure 4.19: Flows and Discharge measurement done downstream of Kafue River confluence.

4.1.3.7   Correlation between measured discharge and reservoir outflows

Figure 4.20 shows a strong positive correlation between the measured discharges and the outflows from the reservoirs, giving $R^2 = 0.997$. This gives enough confidence that the measurements can be used in the water balance calculation. The observed losses can be partially attributed to systematic errors in discharge measuring process and equipment. However, some losses can also be attributed to irrigation, river channel open water surface evaporation and filling up flood plain morphological features and recharge of flood plains' ground water requirement.

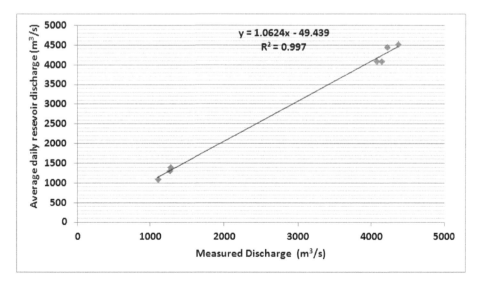

Figure 4.20: Correlation between reservoir outflows and measured discharge.

4.1.3.8   Estimation of sub-catchment runoff from direct catchment rainfall

The estimation of sub-catchment runoff yield is based on the Runoff Curve Number (Soil Conservation Service, 1954). The Runoff Curve Number (RCN) is a coefficient that deduces the total precipitation to runoff potential after losses due to evaporation, absorption, transpiration and surface storage. To get an appropriate RCN, consideration was made of the type of land cover, quality of land cover and soil type. A RCN of 40 was obtained from the table of options provided by the Soil Conservation Service (1986). The methodology has been utilised to obtain an estimate of the local catchment contribution to discharge generation. Four rainfall stations were used for the runoff estimation. The procedure for the calculation is outlined in Table 4.5. The rainy days were grouped in decades for the assessment of the runoff. The total period of assessment is 100 days, from December 2010 to 10th March 2011. A local sub-catchment discharge contribution of 31 m^3/s has been obtained. The month of November 2010, being the start of the rain season had poor rainfall distribution which could not amount to any catchment runoff.

Table 4.5: Estimation of local catchment contribution from rainfall.

10th March, 2011					
Station	Total rainfall (mm)	Catchment area (m^2)	Runoff yield (m)	Runoff (m^3)	Estimated Discharge (m3/s)
Kariba	642.0	4520250000	0.0062	28025550	3.243697917
Mana pools	495.1	5953500000	0.0051	30362850	3.51421875
Lusaka City Airport	748.4	10360500000	0.0157	162659850	18.82637153
Kasaka	702.1	2866500000	0.0165	47297250	5.47421875
					31.05850694

4.1.3.9   Floodplain morphology as a source of water losses

Each morphological feature has its own unique way of either contributing to increasing or reducing discharge in the river reach under review. The discussion below provides an explanation on the possible contribution to increase in discharge losses, which in the water balance study has been designated as sub-catchment storage (ΔS).

Using Google Earth images, a profile is drawn across the flood plain, which helps appreciating the different morphological features of the Middle Zambezi (Figures 4.21 and 4.22). Table 4.6 outlines the floodplain features and classification based on the research observation work done during the field campaigns

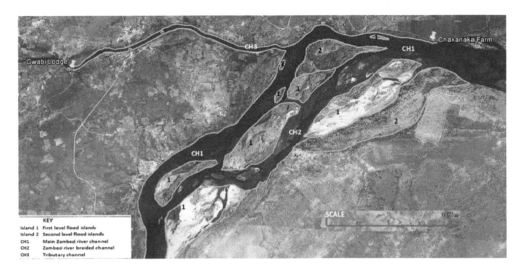

Figure 4.21: Middle Zambezi flood plain morphological features.

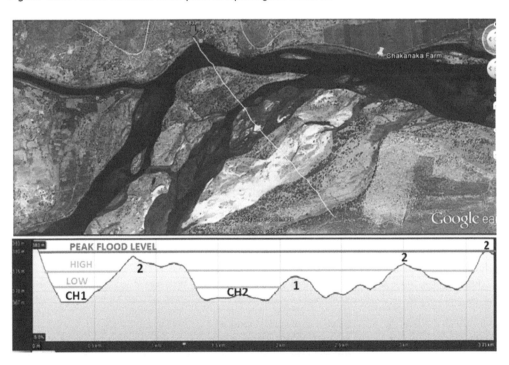

Figure 4.22: A floodplain profile with water levels and morphological features (3.73 km).

Table 4.6: Description of the map and flood plain profile symbols in Figures 4.21 and 4.22.

Symbol	Name	Description	Relation to flooding
1	First level islands	Mainly sand bars with minimal vegetation	Remain as islands throughout regular low flows periods. Subject to changes in size and position after each flood period.
2	Second level islands	With a thin top soil layer of sandy loams but made up of sandy strata	Remain as island at high flood levels. More stable in both size and position after flood events.
CH1	Main Zambezi River channel	Represents that deepest part of the channels preferred for navigation	Will have water even at very low flows
CH2	Alternate braided channels	Some of the channels always have water, while some are intermittent at very low flows	As flows increase, most of the braids become filled and flow, however as flows get to high flood, the multiplicity of the channels decrease tending to all being CH1
CH3	Tributary channels	Represents all tributary streams flowing into the main Zambezi channel, in this sub-catchment the largest tributary being Kafue river	At low flows the tributary channel flow normally to contribute to CH1, however as flows of the main river increase the Zambezi water flows into the CH3 channels up to some distance which would result into increased channel loses
Low	500 - 2,000 m^3/s	Flows mainly confined to the main channel CH1 and braided channel CH2, however some braids become shallower and dry out as flows reduce.	With hydropower water regulation, this flow represents the regular turbine discharge from Kariba (Q1) and Kafue (Q2).
High	2,000 - 4,500 m^3/s	River channels both CH1 and CH2 become wider as islands (1) get submerged, some old channel braids become active at this level of flooding. at this level, flows from main Zambezi River tend to fill up tributary channels CH3 b	Such flows expected with two Kariba spillway gates open in addition to the turbine discharge from Q1 and Q2.
Peak	4, 500 m^3/s and above	River channels CH1 and CH2 merge as all islands are submerged. There is increased reverse flow into the tributary channels CH3 causing extensive flooding along the tributary streams	Wide spread flooding expected when more than two Kariba spillway gates are open in addition to the turbine discharge from both Q1 and Q2

4.1.3.10   Storage losses related to ground water recharge
At the river reach scale, the channel banks and bed tend to lose water to the ground when the channel water level is higher than the underlying water table (Hoehn, 1998). The channel bed and island surfaces can be considered to be surfaces through which main stream water infiltrates to the subsurface allowing for recharge of groundwater. As observed, the island marked (1) in Figure 4.21, is mainly made up of sand bars, which are highly permeable. The islands marked (2) have a thin layer of sandy loams and the strata are mainly sand, which make them highly permeable. Considering that the floodplains are extensive in the middle Zambezi sub-catchment, the loss to the ground water through the banks and river bed would contribute significantly to the increase in sub-catchment water storage (ΔS). However the floodplains geological formation mainly has a perched subsurface aquifer which, when saturated, can contribute to increased discharge into the main channel (CH1) and channel braids (CH2). Section 4.3 describes the study of the subsurface flows, confirming this hypothesis.

4.1.3.11   Water abstraction for irrigation
While most of the study area is protected, there are some open areas which are available for agricultural development. The most commonly grown crop in the Zambezi Valley is banana. This has been found to fare well, given the abundant availability of water and the high temperatures. Figure 4.23 shows a Google Earth image with some farm areas under irrigation in the study area. Figure 4.24 shows a banana field and a pumping station at one of the farms along the Kafue River.
The water abstraction was estimated based on banana water requirement per week, as recommended by Akehurst et al 2008. For the good management of the banana crop, an irrigation requirement of 60 mm of water per week has been specified. The next step was to calculate the total irrigated land. Google earth image was used to compile the farms and calculate the farm areas under irrigation. The total irrigated area is 10,557,849 m². With the irrigation water requirement of 60 mm per week, for successful irrigation of the Banana crop, this translates to a total water abstraction requirement of 633.470.95 m³ per week which is equivalent to 1.05 m³/s. Table 4.7 shows the procedure used in the calculation. This is a negligible value compared to the differences between measured and outflow discharge.
The water balance for both November 2010 and March 2011 therefore includes the component of Q_{irr} which is estimated at 1.05 m³/s.

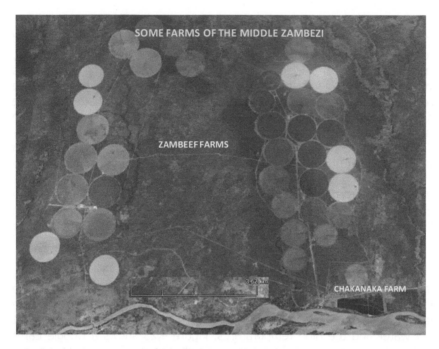

Figure 4.23: Some farms of Middle Zambezi.

Figure 4.24: Banana field and water pumping station at one of the Banana Farms on the Kafue River.

Table 4.7: Calculation of water abstraction.

IRRIGATION PATCHES OF LAND					
No Farms	Area in m^2			Volume per week in m^3	Volume in m^3/s
1 Farm 1 on Kafue River -	374,702.93				
2 Farm 2 on Kafue River	280,625.48				
3 Farm 3 on Kafue River	31,330.00				
4 Zambeef Farm on Kafue River	8,136,724.94				
5 Chakanaka Farm on Zambezi River	1,271,920.95				
6 Farm 4 on Zambezi River	462,544.70				
	10,557,849.00	0.06	633470.94	633,470.94	1.05

Irrigation requirement for bananas is 60mm (0.06m)per week

4.1.3.12   *River channel surface water evaporation*

Evaporation is an important component of the water balance equation. However, because of its nature, surface water evaporation is rarely measured directly, except over relatively small spatial temporal scales (Jones 1992). Evaporation from the water surface is most commonly computed directly by a number of techniques and the most commonly used techniques include: pan coefficient multiplied by measured pan evaporation; water balance; energy balance; mass transfer and; a combination of these techniques. The selection of the technique to be used in a particular situation is largely a function of data availability, type or size of water body and the required accuracy of estimated evaporation values (Jensen 2010). The most common method for estimating evaporation is to use measured evaporation rates from a standard pan and then multiply by a specific coefficient giving rise to the equation 4.3 as follows:

$$E = K_p * E_{pan}$$
$$(4.3)$$

where

K_p is the pan coefficient
E_{pan} is the evaporation rate measured from a class A pan   [mm/day]

Evaporation data used for the computation is from a station within the Kafue catchment at Itezhi-tezhi.
The river surface areas were first measured using the Google Earth images. Distances and channel widths were measured at regular intervals. The average widths of specific river segments are listed in Table 4.8. Then the Class A Pan evaporation data for the month of November was derived. The evaporation data for March 2011, which is within the rain season is negative and therefore was not applied to the data set. Table 4.9 shows the application of the Class A Pan evaporation data application to the surface areas. An evaporation coefficient of 0.75 was applied to the data, to compensate for the overestimations that occur in the pan evaporation.

The following procedure was used to calculate river surface water evaporation:
- Sum the average daily pan evaporation (mm) for the Months of November 2010 and March 2011;
- Multiply the monthly sum of daily evaporation by the Class A pan evaporation coefficient of 0.75(dimensionless);
- Divide the result in (2) by 1000 to yield units of metres;
- Multiply the result in (3) by the open-water area in metres to yield the monthly open-water evaporation in m^3 per month;
- Divide the result in step 4 by 2,592,000 seconds to obtain an evaporation rate in m^3/s

From this procedure an evaporation rate of 3.87 m^3/s was calculated for the month of November 2010. This has been applied to both months of November and March due to the fact that the area is prone to droughts requiring irrigation throughout the year except for exceptionally wet years. This loss is negligible compared to the difference between measured and outflow discharge.

Table 4.8: River segments surface areas.

No	River reach section	Average width	Distance in meters	Distance in km	Surface area (m^2)
1	Kariba to Namoomba	238.66	26,560.00	26.56	6,338,809.60
2	Namoomba to Chirundu	636.56	39,840.00	39.84	25,360,550.40
3	Chirundu to Chakanaka	743.8	17,810.00	17.81	13,247,078.00
4	Chakanaka to Samango	948.01	37,910.00	37.91	35,939,059.10
5	Samango to Chiawa Camp	1,480.30	19,880.00	19.88	29,428,364.00
6	KG Dam to Kafue Confluence	115.8	73,140.00	73.14	8,469,612.00

Table 4.9: Evaporation calculation for the surface area.

Class A Pan Evaporation in mm	Class A Pan Evaporation Coefficient of 0.75	Pan A Evaporation in metres	Estimated evaporation from the river surface area in m^3 per month	Total Evaporation for the month of November in m^3	Total Evaporation for November of m^3/s
112.52	84.39	0.08439	534,932.14	534,932.14	
112.52	84.39	0.08439	2,140,176.85	2,140,176.85	
112.52	84.39	0.08439	1,117,920.91	1,117,920.91	
112.52	84.39	0.08439	3,032,897.20	3,032,897.20	
112.52	84.39	0.08439	2,483,459.64	2,483,459.64	
112.52	84.39	0.08439	714,750.56	714,750.56	
				10,024,137.29	3.87

4.1.3.13 Water Balance calculation

For this study the water balance equation (4.4) has been adopted, here repeated for convenience:

$$Q_{I(1)} + Q_{I(2)} + P - E - Q_{irr} - \Delta S = Q_O$$
$$(4.4)$$

In which

P is the total input due to precipitation from the catchment - this is here treated as the contribution of the local catchment tributaries [m³/s].

$Q_{I(1)}$ is the discharge outflow from Kariba reservoir, treated as discharge inflow (1) into the system [m³/s].

$Q_{I(2)}$ is the discharge outflow from Kafue Gorge reservoir, treated as inflow (2) into the system [m³/s]

E is total loss due to evaporation from the river channel water surface [m³/s].

ΔS is the total loss due to storage within the river reach and flood plain [m³/s].

Q_O is the discharge outflow at the last monitoring station [m³/s].

Q_{irr} is the total water abstraction for irrigation [m³/s].

ΔS has been retained in the equation to take care of the loses. With further studies, the ΔS will be apportioned appropriately to components like evaporation and flood plain groundwater recharge.

Table 4.10: Water Balance equation components for November 2010.

Symbol	Specification	Source of data	Measured	Calculated / modeled	Amount (m³/s)
$Q_{I(1)}$	Discharge inflow (1) into the system from Kariba	Zambezi River Authority			**1,097**
$Q_{I(2)}$	inflow (2) into the system from Kafue Gorge	ZESCO			**206**
P	Runoff from local sub-catchment area (contribution from tributary streams)			√	**0**
E	River reach channel surface evaporation			√	**3.87**
Q_{irr}	Water abstraction for irrigation of bananas			√	**1.05**
ΔS	storage within the river reach and flood plain			√	**34.08**
Q_O	Discharge outflow at the last monitoring station		√		**1,264**

Table 4.11: Water Balance equation components for March 2011 (Flood condition)

Symbol	Specification	Source of data	Measured	Calculated / modeled	Amount (m³/s)
$Q_{I(1)}$	Discharge inflow (1) into the system from Kariba	Zambezi River Authority			**4,153**
$Q_{I(2)}$	inflow (2) into the system from Kafue Gorge	ZESCO			**366**
P	Runoff from local sub-catchment area (contribution from tributary streams)			√	**31**
E	River reach channel surface evaporation			√	**0**
Q_{irr}	Water abstraction for irrigation of bananas			√	**1**
ΔS	storage within the river reach and flood plain and tributary river channels			√	**185**
Q_O	Discharge outflow at the last monitoring station		√		**4,364**

When considering the large upper catchment areas controlled by Kariba (663,000km²) and Kafue Gorge reservoir (152,810km²), with the strong positive correlation between reservoir releases and the flows in the river reach, it can be said that this sub-catchment (23,700 km²) almost entirely depends on the outflows from the reservoirs for its sustainability. However, though the local Middle Zambezi catchment is small representing only 3 percent of the total upper Zambezi and Kafue catchment, it contributes about 1 percent of the total inflows into the Middle Zambezi. The months for which water balance has been calculated do not show any significant contribution of the local catchments to the flows. Although the Middle Zambezi sub-catchment lies within a low rainfall area, with mean annual rainfall of 650 mm, characterised by a short rain season from December to mid March and prone to droughts, it would be expected that in the months when the local catchment area experiences a good rainfall distribution, the local tributaries would be able to make some significant contribution to discharge generation within the catchment.

The study area can be said to be a hydropower dominated river reach and therefore the dam operators should be mindful on how responsive the river system is to the reservoir releases made. The impact of Kariba flood releases contribution to tributary channels should particularly be taken with caution as this has potential to result in flood disasters if both Q1 and Q2 decide to spill within the same period of time.

The flood plain morphological features have their own water needs to fill up the low flow tributary channels and recharge flood plain subsurface ground water, there is a clear loss recorded for the water released from the reservoirs. This water demand presents a significant contribution to the observed discharge losses in the study area. However, other sources of negligible discharge losses include river surface evaporation (E) and water abstraction for irrigation (Q_{irr}).

From the water balance calculation for dry period (November 2010), it shows that 2.6 percent of the total outflows from the reservoir is apportioned to the local catchment storage (ΔS). For the flood period (March 2010) the magnitude of local catchment storage increases to 4 percent of the total outflows from the reservoirs and local catchment contribution.

The water balance studies for both the dry period and the flood period show that the Middle Zambezi River reach is highly dependent on the storage outflows from the hydropower reservoirs. Therefore there is strong need for dam operators to take into account the needs and response of this sub-catchment downstream of the reservoirs.

4.2   Hydrodynamic model simulations

A comprehensive evaluation and characterization of the flow regime before and after the construction of dams is essential to provide measurement of their impact. The hydrodynamic model therefore provides a valuable tool for flow regime alterations evaluation (Matos et al., 2010, Poff et al., 2010).

4.2.1 Model description
To understand the nature and variation of flows in the river reach a model based on the SOBEK-Rural software (www.deltares.nl) was used. This is a one-dimensional open-channel dynamic numerical modelling system for water flow in open channels that has been applied on various river systems all over the world (Prinsen and Becker, 2011). The mathematical model is based on the de Saint-Venant equations for unsteady flow, as outlined in equations 4.5 and 4.6.

$$\frac{\partial A}{\partial t} + \frac{\partial Q}{\partial x} = 0 \qquad\qquad \text{(continuity equation)}$$
$$(4.5)$$

$$\frac{\partial Q}{\partial t} + \frac{\partial}{\partial x}\left[\alpha \frac{Q^2}{A}\right] + gA\frac{\partial h}{\partial x} + \frac{gQ|Q|}{C^2 \mathrm{RA}} = 0 \qquad \text{(momentum equation)}$$
$$(4.6)$$

Where:
A = the wet cross-sectional surface   [m²];
Q = the discharge   [m³/s]
t = time   [s]
x = the longitudinal distance   [m]
α = the Boussinesq coefficient   [-]

g = the acceleration due to gravity [m/s^2]
h = the water depth [m]
C = Chézy coefficient [m$^{1/2}$/s];
R = the hydraulic radius [m]

The model computes the Chézy coefficient as a function of the Mannings coefficient according to:

$$C = k_s \, R^{\frac{1}{6}}$$

where

k_s = manning coefficient [s/m$^{1/3}$]
C = Chézy coefficient in Equation 4.6 [m$^{1/2}$/s]
R = hydraulic radius P/A (P = wet perimeter A = wet area [m]

The following assumptions are used in SOBEK-Rural:

1)   The Model datum and reference level both refer to a horizontal plane from which elevations are defined. All levels are defined with respect to the model datum or reference level.

2)   The bed level is defined as the lowest point in the cross section and is given relative to a reference level, such as Mean Sea Level. Each cross section is definition by interpolation and extrapolation of the bed levels over a reach.

3)   The water depth is the distance between the water level and the bed level.

4)   The water level is the level of the water surface relative to the reference level or datum. The water level is perpendicular to the flow direction and is assumed to be horizontal.

5)   The average flow velocity in the flow through a composite cross section segmented to include the main channel and the floodplain.

The water levels and discharges through the SOBEK-Rural network are calculated using a numerical scheme that solves the de Saint-Venant equations (continuity and momentum equations) by means of a grid in which the water levels are defined at the connection nodes, while the discharges are calculated and defined at the reach segments.

4.2.2 Model setup and scenarios
The upstream and downstream boundaries of the model domain are: the Kariba Gorge cross-section (upstream boundary); the Mupata Gorge cross-section (downstream boundary). Tributary 1 was defined as the Kafue Gorge Dam. Figure 4.25 shows the Middle Zambezi SOBEK network and the model domain.

A total of 53 floodplain/channel cross-sections were reconstructed using the Google Earth SRTM DEM, while 6 cross-sections were constructed on the tributary 1 (Kafue). The correction and reconstruction of the cross-sections were done by using the 5 field-measured cross-sections, but at least one cross-section had to be constructed every 3 km in the river reach. Figure 4.26 shows the river channel cross-section reconstruction process. Each reconstructed cross-section was assigned a specific Manning's roughness, as shown in Figure 4.27. The process of definition of the cross-sections led to the definition of the logitudinal Middle Zambezi River reach profile. To validate the model generated profile, the ADCP measured river reach longitudinal profile was superimposed. A number of bed morphology features were identifiable and compared on both profiles, Figure 4.28 shows the superimposed profile.

The generation of network water levels through the nodes and generation of discharge through the reaches are shown in Figures 4.29 and 4.30.

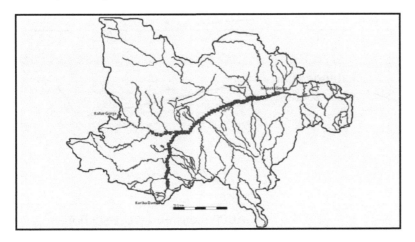

Figure 4.25: Middle Zambezi SOBEK network.

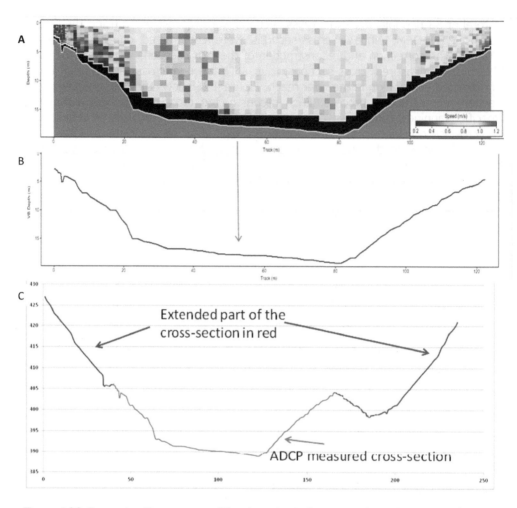

Figure 4.26: Reconstruction process of the river channel cross-sections. A: measured wet cross-section with ADCP. B: contour of the measured cross-section. C: extended cross-section including floodplain topography (transverse profile derived from Google earth).

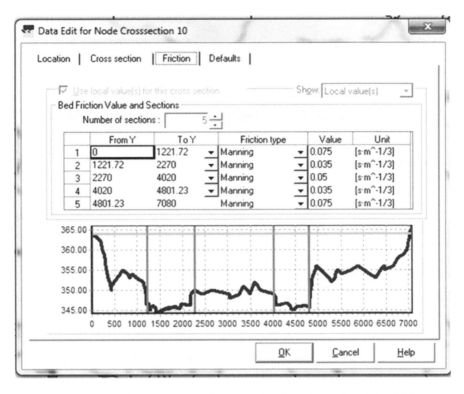

Figure 4.27: Process of assigning roughness to the river channel and floodplain cross-sections.

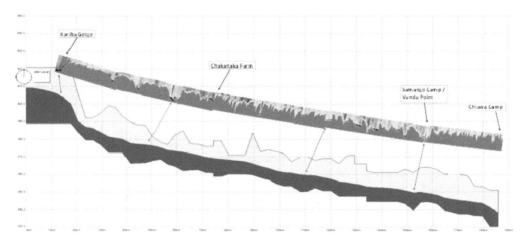

Figure 4.28: Model generated profile of the Middle Zambezi at Q = 7,000 m³/s - superimposed with the field ADCP measured flood flow longitudinal profile.

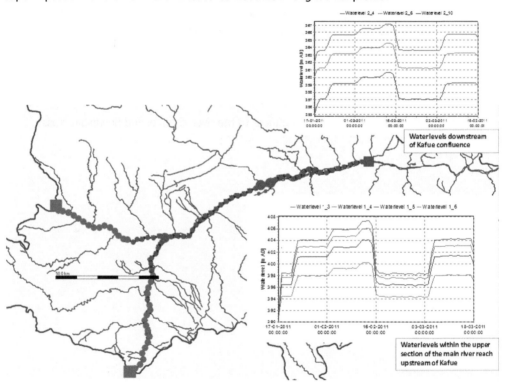

Figure 4.29: Water level calculation through the nodes.

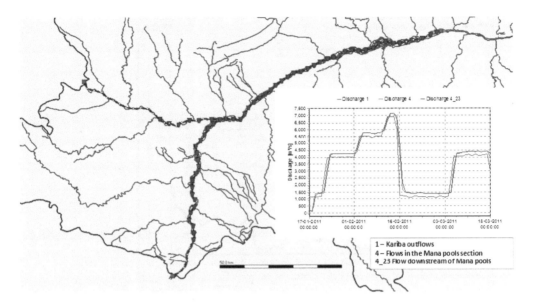

Figure 4.30: Discharge generation through the reaches.

4.2.2.1   Model calibration based on water levels

The diver-measured water levels were used to optimize the value of the channel roughness. The observation period from 28th November 2010 to 17th May 2011 was compared to the model generated water levels using a range of Manning values between 0.03 s/m$^{1/3}$ and 0.05 s/m$^{1/3}$, the model tends to overestimate the water levels for the low flow periods of the dry season, but has a good fit for both high flows and low flows of the wet season, as can be seen from Figure 4.31. This range of roughness provided the best fit between the simulation and the measured water levels (Figure 4.32).

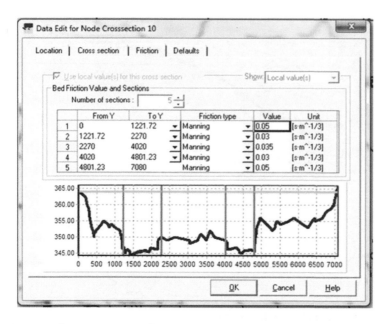

Figure 4.31: Manning roughness cross-section assignment for ZM5.

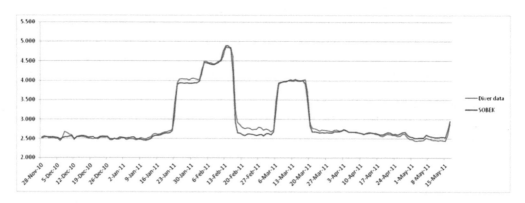

Figure 4.32: Calibration using Manning's roughness (0.03 to 0.05) using diver-measured water levels.

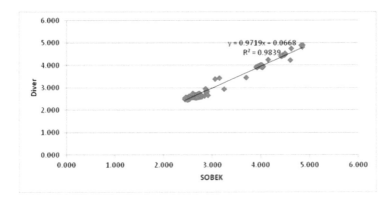

Figure 4.33: Comparison plot - R^2 = 0.9839.

4.2.2.2 Model validation based on discharges

To check the performance of the model, the data set of field measured discharge was used (Figures 4.28 and 4.29). The model performed well in predicting the measured discharges for both low flows and high flows (Figures 34 and 35).

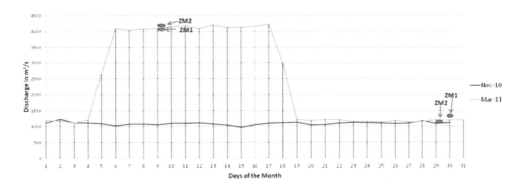

Figure 4.34: Model validation using field discharge measured data.

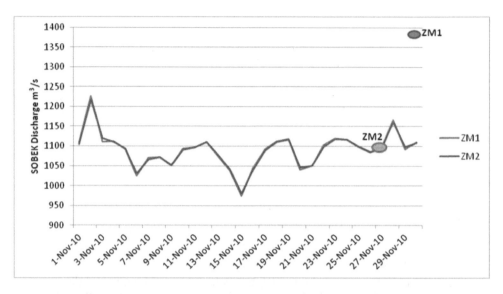

Figure 4.35: Model validation for low flows using field measured discharge data.

4.2.3 Comparison between the past and current surface flows states

4.2.3.1   Scenario without dams

The scenario without dams was simulated using unregulated reservoir inflows from the Zambezi River at Victoria Falls and the Kafue River at Kafue Hook Bridge, these being the upstream stations not affected by any form of regulation (Figure 4.36). The analysis of the results of this simulation shown in Figure 4.36 allows observing the following:

a) Predictable natural rhythm of annual increase and recession of flows.
b) Minimum low flows depending on the catchment rainfall for that particular hydrological year.
c) Floods occurring at irregular intervals of years with high annual peaks flows.
d) Kafue system contributing significantly to peak floods, note the green hydrograph in Figure 4.36.
e) Dry period observed from late 1970s to early 2000s, with 11 years of less than 2,500 m^3/s peak flows.

The water levels show regular floodplain flooding, as indicated by Table 4.12 showing the wet surface width as a function of discharge.

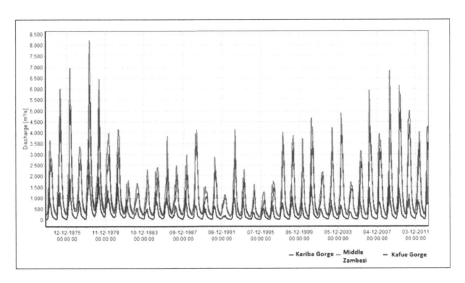

Figure 4.36: Unregulated Middle Zambezi River reach discharge simulation - natural rhythm of peaking and recession for both the extremely wet and dry periods.

4.2.3.2   Scenario with dams

This simulation used the regulated flows from Kariba Q1 and Kafue Gorge Q2. Kariba levels show two dry years (1987/1988 and 1995/1996) in which the Kariba reservoir level was exceptionally low. In those years no flood spillage were done and two wet years (2011/2012 and 2012/2013) when reservoir levels when slightly above the reservoir operating rule curve, that calls for spillage (Figure 4.31). The results are shown in Figure 3.37. The key features are:

a) An increase in minimum low flows, these are showing an upward trend to not less than 1000 m^3/s from the year 2001 – may be an indication of the up rating of the power stations' generation machines and increased electricity demand.
b) Lack of flood spillage for extended periods, may be due to low reservoir inflows leading to low reservoir levels.
c) Flood spills only occurring when reservoirs are full.

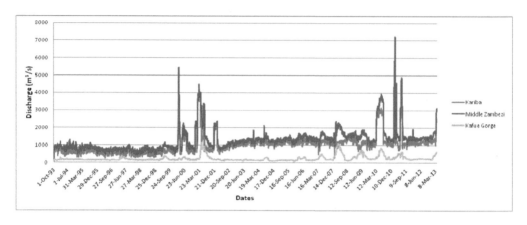

Figure 4.37: Scenario with dams.

4.2.3.3 Comparison between scenarios

Figures 4.38 and 4.39 allow comparing the unregulated and the regulated regimes, representing the past and the present situations, respectively. Table 4.12 shows the analysis of comparison in terms of floodplain wetted area for the different flow regimes (minimum, moderate and maximum) for both the unregulated and regulated situations. Figure 4.40 shows the wet area extension at Chakanaka (ZM3) as a function of the river discharge.

The following can be deduced:

a) The predictable rhythm of recharge and recession is lost.
b) The seasonal low flows have become higher, being about 1,000 m^3/s up to 2001, and above 1,000 m^3/s and rising for the period from 2001. The increase can be attributed to the full operation of all the three power stations and a general increase in electricity demand in the region. In the natural system, the low flows would be less that 100 m^3/s in the dry season. Low seasonal flows have therefore increased from 100 m^3/s to over 1000 m^3/s.
c) Scarce and irregular flood spillages, with 4 spillage flows in a period of 22 years (1993 to 2013).
d) Short duration of the flood spills, some lasting only a few weeks.

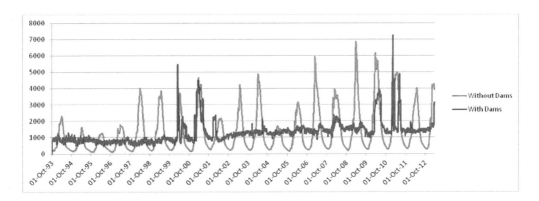

Figure 4.38: Discharge comparison between the two scenarios without dams and with dams (values in m^3/s).

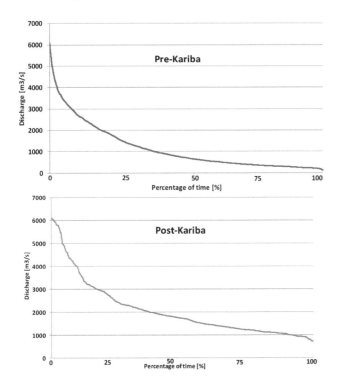

Figure 4.39: Pre-Kariba and Post-Kariba flow duration curves showing an increase in the low flows through the river reach.

Table 4.12: Comparison of floodplain flooding between the unregulated and regulated flows (minimum, moderate and maximum).

Discharge	Cross-section - Chakanaka Farm		Cross-section - Chiawa Camp		Cross-section 10 Chikwenya island	
	Water level (masl)	Floodplain width (m)	Water level (masl)	Floodplain width (m)	Water level (masl)	Floodplain width (m)
Q_1_Unregulated	367.5	777.0	350.5	1661.1	346.0	1197.6
Q_2_Unregulated	368.5	860.0	351.5	1762.5	347.0	1762.0
Q_{max}_unregulated	369.5	1062.5	352.5	1995.0	348.0	1804.2
Q_1_regulated	366.7	744.0	351.0	1673.0	348.6	1859.3
Q_2_regulated	367.9	854.0	351.9	1764.0	349.1	2679.2
Q_{max}_regulated	368.7	884.5	352.5	1995.0	349.6	3233.0

Where
Q_1_Unregulated = 4,500 m^3/s
Q_2_Unregulated = 6,500 m^3/s
Q_{max}_unregulated = 8,500 m^3/s

Q_1_regulated =4,500 m^3/s (two spillway gates open)
Q_2_regulated = 5,800 m^3/s (3 spillway gates open)
Q_{max}_regulated = 7,000 m^3/s (4 spillway gates open)

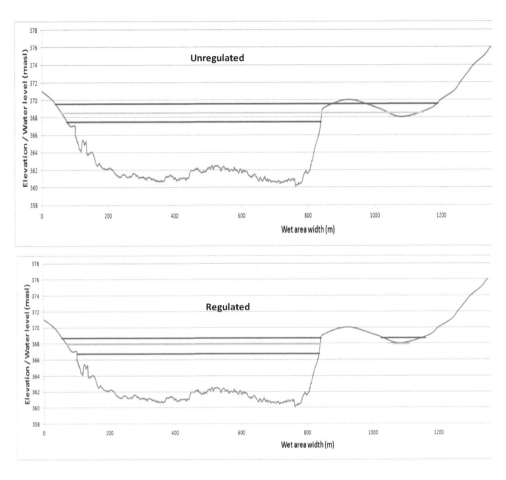

Figure 4.40: Cross-section 35 - Zambezi at Chakanaka (ZM3) area showing the wet area extension for Qmax, Qmoderate (Q2) and Qminimum (Q1) - as stated in Table 4.12.

4.3 Subsurface flow model simulations

This part of the study was carried out within the framework of an MSc project co-supervised by the author of this work (Khan, 2013, published in Khan et al., 2014). Groundwater modelling was carried out to study the interaction between the groundwater and surface flows, and how this interaction is affected by the changes in the flow regime and in turn affects the river morphology.

4.3.1 Model description

The investigation is based on the application of Processing Modflow (PM). This is a computer code accompanied by a graphical user interface (GUI) developed by the U.S. Geological Survey. PM consists of a modular finite-difference model for flow in a porous medium. It is used to solve the groundwater flow equation to simulate the flow in aquifers. For the Middle Zambezi River floodplain, the main objective was to model the groundwater-surface water interaction. For this, the additional modules RIVER and RECHARGE were used. The governing equation for PM uses finite-difference methods to solve the 3D groundwater flow equation in a saturated heterogeneous and anisotropic porous medium, which is given by:

$$\frac{\partial}{\partial x}\left[K_{xx}\frac{\partial h}{\partial x}\right] + \frac{\partial}{\partial y}\left[K_{yy}\frac{\partial h}{\partial y}\right] + \frac{\partial}{\partial z}\left[K_{zz}\frac{\partial h}{\partial z}\right] + W = S_s\frac{\partial h}{\partial t} \qquad (4.7)$$

In which:

K_{xx}, K_{yy}, K_{zz} values of hydraulic conductivity along x, y and z axes
    [m/s]

h total head   [m]

W sources and sinks, in terms of volumetric fluxes   [s^{-1}]

S_s specific storage   [m^{-1}]

t time   [s]

To solve the equation, PM uses the finite-difference method where the aquifer domain is segregated using a model grid. The grid is organized in rows, columns and layers (i, j, k) as outlined in Figure 4.41

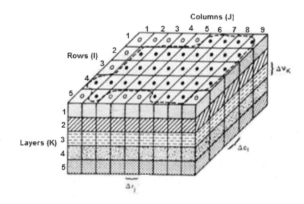

Figure 4.41 - Example of a model grid. Source: Zhou (2012).

The following assumptions are used in MODFLOW:
- There are no non-orthogonal anisotropies in the model area (as, for instance, because of a fracture in a rock);

- The water has a constant density and dynamic viscosity over the model domain.

These assumptions are close to the state of the study area since the aquifer is in alluvial deposits where no fractures are present. The constant density and viscosity form a reasonable assumption for this case as well.

The river package was used to simulate the flow between the river (surface water) and the surrounding aquifer and analyze the interaction between the groundwater and surface water under different conditions. In this package the following parameters were defined (using the polyline input method):

- Hydraulic conductivity of river bed;
- Head in the river;
- Elevation of the river bottom;
- Width of the river;
- Thickness of river bed sediments;

The flow rate between river and groundwater was computed with the following equation:

$$Q_{riv} = C_{riv} \cdot (h_{riv} - h) \qquad \text{if } h > B_{riv} \tag{4.8}$$

$$Q_{riv} = C_{riv} \cdot (h_{riv} - B_{riv}) \qquad \text{if } h < B_{riv} \tag{4.9}$$

In which:

Q_{riv}	flow rate between river and groundwater	[m^3/day]
C_{riv}	hydraulic conductance of river bed	[m^2/day]
B_{riv}	elevation of the bottom of the river bed	[m]
h_{riv}	head in the river	[m]
h	head in the aquifer	[m]

The hydraulic conductance is computed by applying the following equation:

$$C_{riv} = \frac{K_{riv} \cdot L \cdot W_{riv}}{M_{riv}} \tag{4.10}$$

In which:

K_{riv}	hydraulic conductivity of the river bed	[m/day]
L	length of the river within a cell	[m]
W_{riv}	width of the river	[m]
M_{riv}	thickness of the river bed	[m]

The recharge package was used to allow inputting the recharge fluxes to the groundwater. The value of the recharge to the groundwater depends upon the water balance in the root zone, where precipitation and evaporation or evapotranspiration play a major role. As information regarding these parameters was not available during the research period to do the root zone water balance, the recharge value was chosen based on experience from other studies and then calibrated using the PEST (parameter estimation) model in Modflow. The model uses the input - recharge flux (I_R) to compute the recharge rate (Q_R) by applying the equation below:

$$Q_R = I_R \cdot DELR \cdot DELC \qquad\qquad (4.11)$$

In which: $DELR$ and $DELC$ are the sizes of the grid cell sides in row and column respectively, and their multiplication gives the area of the cell.

4.3.2 Model setup

The setup of the model, defined parameters and characteristic of materials and other aspects of the model have been presented. The topography of the model area, defined in the model as the top layer, was based on the SRTM 90 metres DEM. The limits of the aquifer were carefully studied based on the geology of the area. Since the model area is located in a rift valley (down-faulted graben) the aquifer was limited from the sides by the Mozambican and the Zambezi mobile belts which are composed by virtually impermeable granite and gneisses.

The model was constructed with a closed domain where the heads (water levels) are defined by the interaction between the aquifer and the river. So the heads on the river upstream and downstream boundary conditions were defined (using the river package).

The geological materials present in the model domain comprise the following:

- Sandstone, siltstones and mudstones from the Karoo formation. These materials are present on the eastern and western sides of the model domain.
- Recent sandy alluvial formation - It is present along the main river in the sections where the river is alluvial (outside of the gorges), which is in the centre of the model domain along the Zambezi River.
- Old sandy alluvium - It is located in the river terraces, in the South of the model;
- Kalahari sands formation - Comprises sandy soils and is located in the centre of the model domain.
- Basement granite and gneisses - These are hard rocks located on the North and south of the model domain that form the boundary of the aquifer.

The parameters representing the characteristics of the geological materials, which were used in the model, are as presented in Table 4.13:

Table 4.13: Model - material properties.

Material	Hydraulic Conductivity [m/day]
Sandstone, siltstone and mudstone (Karoo)	2.8
Recent sandy alluvium	25
Old sandy alluvium	15
Kalahari sands	50

For part of the materials, the values of the hydraulic conductivity were obtained from the report of Interconsult (1985), which presented some data related to field tests carried out for an assessment of groundwater potential. Due to lack of data, the hydraulic conductivity adopted for the rest of the materials was based on experience and recommended values found in hydrogeology literature. The grid size for the model was chosen taking into account the simulation time and the number of cells inside the model domain to give sufficient detail of the groundwater flow. The model domain is around 165 km in the East-West direction (x-direction) and around 100 km in the North-South direction (y-direction). The cell size is 1000 m in the x direction and 500 m in the y direction. As the main flow is oriented in the y direction the cell size was defined to be lower in y to allow more detail in the flow simulation. With the above sizes the grid has 164 cells in x direction and 195 cells in the y direction.

In the absence of a longer record of groundwater levels, the groundwater level measurements for a shorter period of 5 months (from September 2012 to January 2013) on the left bank of the Zambezi River were used for calibration. The calibration was carried out in steady state mode by using the available data, and was executed automatically using the PEST model (Parameter Estimation), which is easily coupled to Modflow and adjusted manually afterwards. In the calibration procedure the recharge values were tuned in order to obtain a match between the observed and computed values. The difference between the computed and observed values is about 11 cm, as shown in Table 4.14, which can be considered quite acceptable.

Although the simplified calibration is one of the major limitations of the groundwater model, it is believed that the model still provides a general idea about the behaviour of the system.

Table 4.14: Calibration result for steady state.

	Observed (Well in floodplain)	Computed (MODFLOW)
Water Levels [m]	347.5	347.61

4.3.3 Current subsurface flow state

Two models were developed to simulate different aspects related to the interaction between groundwater and surface water in the Middle Zambezi. For the first model, two simulations were carried out with the intention of understanding the impact of the river flow regulation on the groundwater levels along the year. The first simulation had the conditions representing the river water levels and bed levels corresponding to the present, whereas the second simulation used the conditions representing the past (pre-Kariba). The analysis was based on the comparison between the results of the two simulations. For the second model two simulations were carried out. These were to verify how the water table recedes after a natural recession in the river water levels as in the Pre-Kariba period (first simulation), and then after a rapid river water level drop as in the Post Kariba period (second simulation). Table 4.15 outlines two set model simulation description and Figure 4.42 shows the scenario recession curves for the pre-Kariba and post-Kariba situation. Figure 4.43 shows simulation results for aquifer hydraulic heads and flow direction.

Table 4.15: Model scenario description.

Model	Simulation	Recharge	River water levels	Hydraulic conductivities
Model 1	Post-Kariba 2	Same	Present conditions	Same
	Pre-Kariba		Past conditions	
Model 2	Post-Kariba 2	Same	Rapid Recession	Same
	Pre-Kariba		Smooth recession	

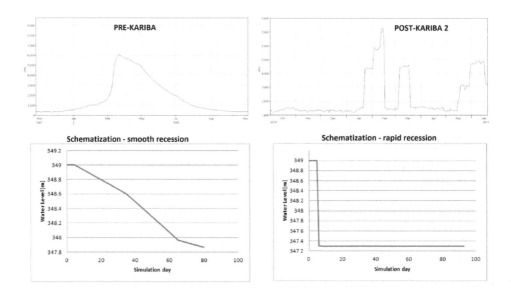

Figure 4.42: Hydrographs considered and their schematisation in the model (Khan, 2013).

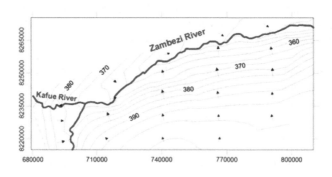

Figure 4.43: Simulation results showing aquifer hydraulic heads. Flow directions are as indicated by the arrows (Khan, 2013).

In the transient model two simulations were carried out, one with a smooth (natural) flood recession rate and another with a rapid recession rate as occurs in the Post-Kariba flow regime. The peak of the hydrograph occurs on simulation day 5 and after this day the discharge drops. The results are presented in Figure 4.44, showing a smooth flood recession rate with the water table remaining almost constant until simulation day 35 and then starting to drop, whereas with a rapid flood recession rate the water table drops sharply, immediately, on the first day already. This reveals an intense flow that occurs from aquifer to the river when the flood recession is rapid which has negative implications on the stability of the sandy banks.

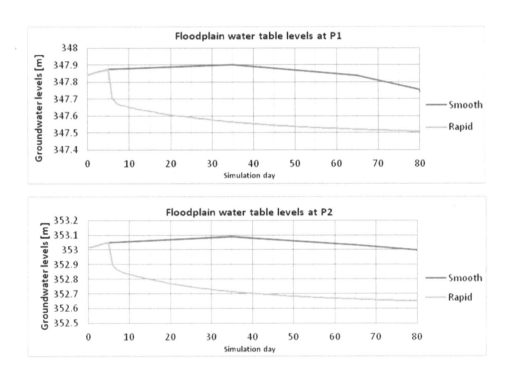

Figure 4.44: Recession of the floodplain water table (Khan, 2013).

The main finding of the steady state simulation is that the regional aquifer drains towards the Zambezi River. This result is supported by some field observations carried out in the framework of this research. Figure 4.45 shows an eroding bank with signs of groundwater outflow towards the river. The discharges from aquifer to river are around 5 to 10 m^3/s, which is an insignificant contribution when compared with the Zambezi average flows of 1,200 m^3/s. However, from the river morphology point of view this is an important aspect, since it has significant consequences for bank erosion. The fact that the aquifer hydraulic heads are higher than the river hydraulic heads means that a rapid water level drawdown would increase significantly the hydraulic gradient in the pores, unbalancing the seepage forces and affecting the stability of the sandy banks of the Zambezi River.

Figure 4.45: Eroding bank with signs of a mottled zone in the lowest layer. The red coloured soil indicates oxidised iron, which is likely to be due to ground water outflow towards the river.

4.4   Conclusions

The water balance of the study river reach for both the dry and the flood period shows that the Middle Zambezi flow is highly dependent on the storage outflows from the hydropower reservoirs. Therefore there is strong need for dam operators to take into account the needs and response of this sub-catchment downstream of the reservoirs.

The impact of the hydropower regulation on the downstream environment led to permanent changes which may not be reversible. The following impacts fall into this category: the water levels significantly vary on the 24-hour circle leading to a constant diurnal water level variation within a range of 0 to 0.2 metres corresponding to the power generation and there has been an increase in dry season flows to over 1,000 m^3/s. This constant flow shows that the study area is hydropower dominated and most of the flows in the river reach are turbine outflows from Kariba (Q1) and Kafue (Q2).

The induced hydraulic effects of the current dam operation can be mitigated to a certain extent: the complete absence of flood flows for extended periods; the extremely short flood spillage periods; and the phenomena of extreme range of 0 to 5 m of water level fluctuation during the opening and closing of flood gates, leading to an extreme departure from the natural system resulting in negative consequences in terms of ecology and morphology. This category of changes can be mitigated by the dam operators and may need an effort in policy shift in the dam operating rules.

The main finding of the steady state sub-surface flow simulation is that the regional aquifer on the floodplains drains towards the Zambezi River, because this enhances river bank erosion. This reinforces the need to make a shift in the way the flood gates are opened and closed to save the downstream river channels and the environment in general.

CHAPTER 5: MORPHODYNAMICS

Part of the contents of this Chapter has been published in Khan, Mwelwa-Mutekenya et al. (2014). This chapter describes the results of historical map and satellite image analysis and morphodynamic modeling which were carried out to assess the morphological changes occurred in the post-impoundment period and to understand the role of the dam, respectively.

5.1 Sediment

5.1.1 Sediment entrainment, transport and deposition processes

Sediment entrainment, transport and deposition processes are key components of the river morphological adaptation. The sediments transported by the river can be classified based on their origin and according to the mechanism of transport, Figure 5.1 shows the sediment transport classification after Jansen et al., (1979). Sediments can be classified based on the two key mechanism by which they are transported, considering the volume of sediments being transported (load), which are: bed load and suspended load. "Bed load" refers to the transport of material by rolling, sliding or saltation on the bed and "suspended" load refers to the transport of sediments in suspension.

Sediments can also be classified according to their origin which are: "bed material load" referring to the transport of sediments that are entrained from the river bed or banks; and "wash load", which refers the material coming from outside the river with negligible interaction with the river bed. This material has particles finer than the sediment found in the river bed and banks. As it can be seen in Figure 5.1, the bed material can be transported as bed load or suspended load, whereas wash load is transported only in suspension.

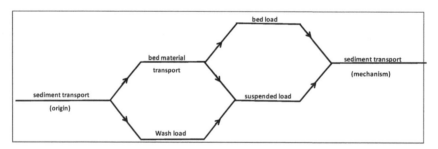

Figure 5.1: Sediment transport classification according to Jansen et al. (1979).

The transported bed material is related to the capacity of the flow to transport the sediment. So it only occurs when the flow velocity is high enough to trigger the initiation of movement. The volume of bed material that is transported by the flow has an upper limit which depends on the flow strength, which is in turn a function of local flow velocity. For this, it is also called capacity-limited transport. Instead, the wash load has almost no relation with the transport capacity of the river. Its load is determined by the quantity of material brought into the river by erosion in the catchment areas upstream. This is why it is also called supply-limited transport.

The construction of a dam can drastically alter the sediment transport in the downstream river, because dams affect both the sediment transport capacity, by altering the water flow, and the sediment supply, by retaining large parts of the sediment in their reservoirs (Kondolf et al., 2014). As a result, also the characteristics of the river bed might change. It might become coarser (Kondolf, 1997) or finer (Ma et al., 2011), depending on the way the dam is operated (Kondolf and Wilcock, 1996).

Kunz (2011) recently estimated the volume of sediment trapped in Lake Kariba and Lake Kafue in about 4×10^6 and 400×10^3 tonnes/year, respectively, but these quantifications present strong uncertainties. The sedimentation rates in Kariba Lake mainly depend on the contribution by tributaries, since the Zambezi River is believed to transport relatively small amounts of sediment after passing through the Barotse and Chobe swamps (Mukono, 1999). For Lake Kafue, the major contribution of sediment is the Kafue River itself (Kunz, 2011). Before dam construction, however, the sediment used to settle in the Kafue flats, which extended to the Kafue Gorge, and in wetlands upstream of Kariba Gorge, so that little changes in sediment inputs to the Middle Zambezi River can be expected from both the Kafue and the Upper Zambezi with respect to the pre-impoundment periods (Khan et al., 2014).

5.1.2 Field sediment data collection and analysis

To get a good understanding of the nature of the sediments transported by the rivers to the Middle Zambezi floodplains and assess the contribution of each tributary, a network of sediment sampling stations was established in the framework of this study. Both suspended and bedload sediment samples were collected at selected locations. Figure 5.2 shows these locations and Table 5.1 shows the station names, coordinates and elevation.

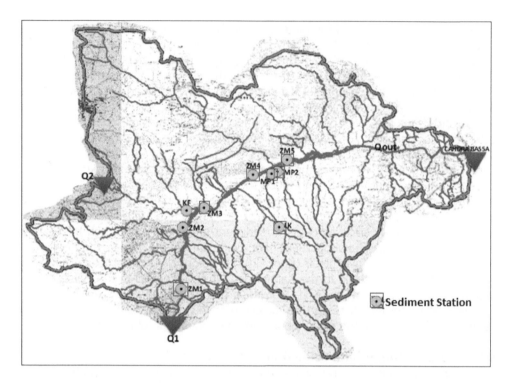

Figure 5.2: Hydrometric network with sediment sampling stations. Red triangles indicate dams, with Q1 being the outflow of Kariba Dam and Q2 being the outflow of Kafue Dam.

5.1.2.1 Suspended sediment

Samples were collected to obtain an indication of the concentration of suspended sediments in the water column during the rainy season. Figure 5.3 (A) and (B) shows suspended solids for both the main river reach and the tributaries. The trend shows a steady increase of suspended solids concentration in downstream direction. The higher values of suspended solids concetnration in the tributary samples (RK1, RK2, MP1 and MP2, see Figure 5.3(A)) indicates that the tributaries have a significant contribution to the total load of suspended solids in the Middle Zambezi River.

Table 5.1: Hydrometric network station names and location.

No.	Name of Station	Latitude (S)	Longitude (E)	Elevation (m asl)	Monitoring Equipment
ZM1	Zambezi River at Namoomba	$16^0 21'01.1''$	$028^0 50'13.0''$	386	Gauge Plates
ZM2	Zambezi River at Chirundu Bridge	$16^0 02'16.3''$	$02851'02.8''$	373	Q measurement only-
KF	Kafue River at Gwabi	$15^0 57'05.5''$	$028^0 51'38.9''$	380	Gauge Plates / Diver
ZM3	Zambezi River at Chakanaka Farm	$15^0 56'20.1''$	$028^0 57'03.1''$	367	Gauge Plates / diver / baro diver
	Zambezi River at Samango Camp	$15^0 45'53.1''$	$029^0 13'20.1''$	367	Gauge Plates / Diver
ZM5	Zambezi at Chiawa Camp	$15^0 41'05.8''$	$029^0 24'47.9''$	357	Gauge Plates / Diver / well Diver
MP1	Mana River at Long Pool	$15^0 44'51.5''$	$029^0 21'20.1''$	359	Diver
MP2	Mana River at Mana bridge	$15^0 44'20.7''$	$029^0 22'04.8''$	359	Diver
LK	Lukumechi river at bridge	$16^0 03'28''$	$29^0 24'28''$	473	sediment
Q1	Kariba Dam	$16^0 31'18''$	$28^0 45'41''$	488	Inflow1(main)
Q2	Kafue Gorge Dam	$15^0 48'30''$	$28^0 25'15''$	994	Inflow2(tributary)
Qout	Mupata Gorge	$15^0 38'13''$	$30^0 01'53''$	337	Out flow

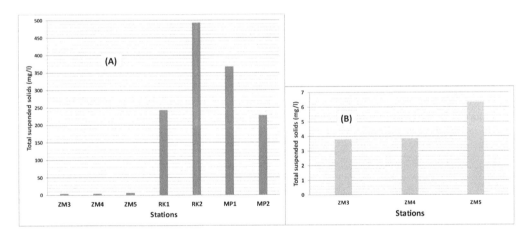

Figure 5.3: Suspended sediment concentrations showing tributary contribution downstream of Station ZM4. (A): Concentrations in main Zambezi and tributaries (RK1, RK2, MP1 and MP1) and (B): Concentrations in the main Zambezi (ZM3, ZM4 and ZM5).

5.1.2.2   Bed sediment
River flow resistance and sediment movement are said to be directly related to the bed sediment size. Therefore, it is important to obtain a good understanding of sediment size distributions in the Middle Zambezi River reach (Vanon, 1975). A

number of bed samples were collected during both the low and high flow periods. These samples were then subjected to laboratory sieve analysis. Figure 5.4 shows the predominant type of sediment that characterizes the main-river channel bed. The size distribution analysis shows D_{50} ranging between 0.15 mm to 1.05 mm (fine to coarse sand), as shown in Figures 5.5 and 5.6. The tributaries tend to have finer bed sediment (0.35 to 05 mm: fine to medium sand) while ZM1, the station immediately downstream of Kariba Dam has the smallest sediment size, the D_{50} being there 0.15 mm. Thereafter, Station ZM3 which is immediately downstream of the Kafue-Zambezi confluence has the largest size (Kafue River may have an influence here). The most downstream station, ZM5, tends to have more silt and clay, which may be due to the contribution from the tributaries, which tend to carry high loads of suspended sediment, as it can be seen in Figure 5.3 (A).

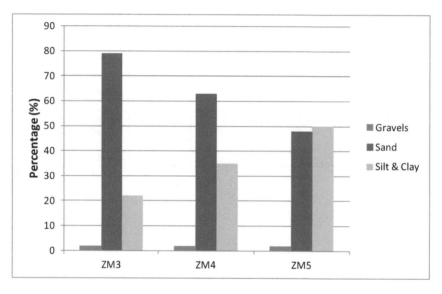

Figure 5.4: Classification of Bed sediments from Middle Zambezi River.

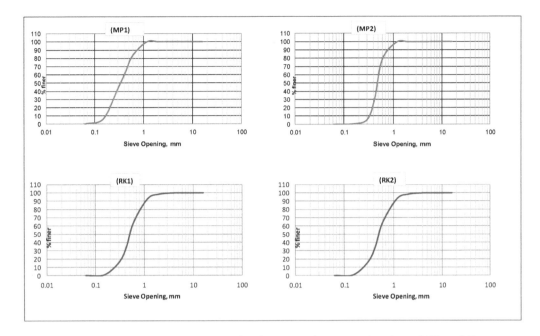

Figure 5.5: Tributary bed sediment size distribution with D_{50} ranging from 0.35 to 0.5 mm. MP1: Mana River at Long pool. MP2: Mana River at Bridge. RK1: Rukomeshi River at Bridge. RK2: Rukomeshi River downstream of Bridge.

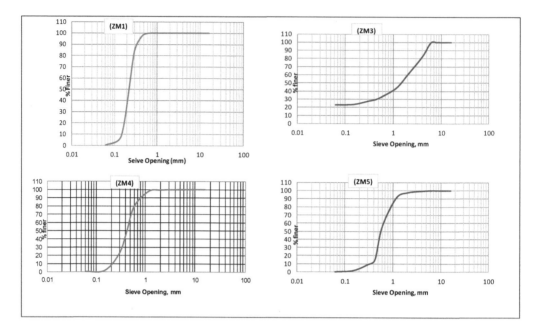

Figure 5.6: Main river reach bed sediment size distribution with D_{50} ranging from 0.15 to 1.05 mm. ZM1: Zambezi at Namoomba (immediately downstream of Kariba). ZM3: Zambezi at Chakanaka (immediately downstream of Kafue River). ZM4: Zambezi at Samango. ZM5: Zambezi at Chiawa.

5.2 Historical morphological changes

5.2.1 Map analysis

The analysis of historical morphological changes at a small scale was carried out on the Chikwenya Island near the Mana pools floodplains. Based on the topographical maps 1958, 1973 and 1977, it becomes evident that the area of bars and islands reduced over time from 7.89 km² to 5.67 km² (Figure 5.7). Such a reduction of bars and islands is linked to the bank erosion phenomenon, already observed by Guy (1981) and observed also during the research period. The sudden closure of the Kariba Dam flood gates (a lack of phased closure of dam flood gates), immediately appeared related to bank collapse during the February 2011 field observation. Four spillway gates were opened between January and February of 2011 and then suddenly closed at midnight of the 16th February 2011. This caused a sudden reduction of flow from 6,944 m³/s to 1,251 m³/s and resulted in extensive bank failure. In some cases, 5 to 10 m of river bank were lost through excessive river bank erosion. A relation between dam operation and bank failure was indicated by du Toit (1983) already. The effects on water level changes of the closure of the dam gates and the worsening effects of ground water flow are discussed in greater detail in Sections 4.1.1.2 and 4.4 of this thesis.

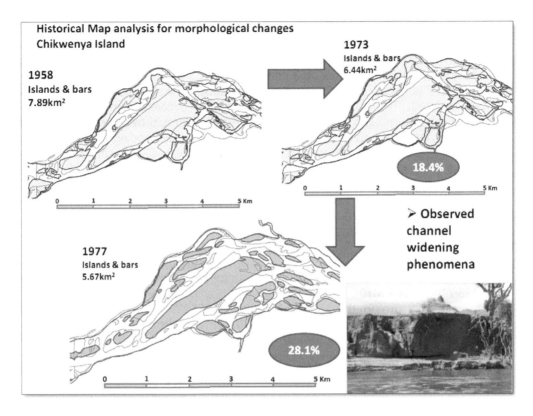

Figure 5.7: Historical map analysis for the Chikwenya island near Mana Pools and Lower Zambezi National Parks.

5.2.2 Satellite image analysis

The changes in channel width were analyzed also using topographical maps and LANDSAT TM imagery in the framework of the MSc. research of Mubambe (2012). Mubambe's work concentrates on 3 channel locations near Mana pools, as it can be seen from Figure 5.8. Table 5.2 outlines the topographical maps and imagery that were analyzed. The analysis shows that the river evolved from a single-thread channel in 1954, as seen from topographical maps to a multi-tread breaded channel. Fluvial activities were enhanced by the dry season artificial flooding due the irregular opening of the dam spill way gates at Kariba before the construction and commissioning of the second power station on the north bank (Kariba North Bank Power station). This occurred during the period referred to as post Kariba 1, which was characterized by annual spillage of water with only one power station (Kariba South Bank Power Station) needing the turbine flows.

The analysis of the results shows that the river channel widened from a width of 1.3 Km (1954) to a width of 2.1 km (2010) at location (A), representing an increase of 61.5% in 56 years (Figure 5.9). At location (B), the river presented the largest widening, from 1.1 Km to 2.9 km in the same period, representing a width increase of 138.7% (Figure 5.10). At location (C), the river presented initial widening, from 1.9 km to 2.8 km in the first operational period (Post-Kariba 1), followed by narrowing to 1.2 km in the period 1976 - 1984. In the period 1984 - 1992 the river width remained constant (1.2 km) to widen again afterwards, reaching the value of 2 km in 2010, representing an increase of width of 5.8% in the period 1954 - 2010 (Figure 5.11). Table 5.3 gives a summary of the measurements at the three locations (A), (B) and (C).

Table 5.2: Images and topographical maps used in the analysis.

No	Image/Map sheet type	Image/Map Sheet Date	Resolution
1	Topographic Map	July 1954	-
2	Topographic Map	July 1976	-
3	Landsat TM Image	July 1984	30 meters
4	Landsat TM Image	July 1992	30 meters
5	Landsat TM Image	May 2010	30 meters

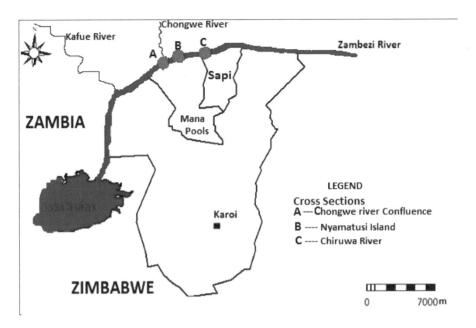

Figure 5.8: Location of TM Satellite analysed cross sections (adapted from Mubambe, 2012).

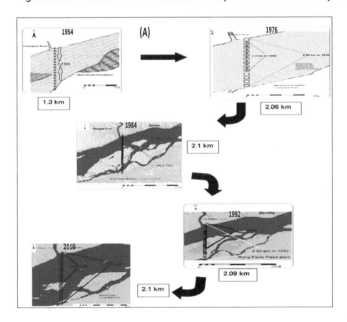

Figure 5.9: Changes in cross section of the Zambezi River near the Zambezi-Chongwe confluence - at location (A) (after Mubambe,2012).

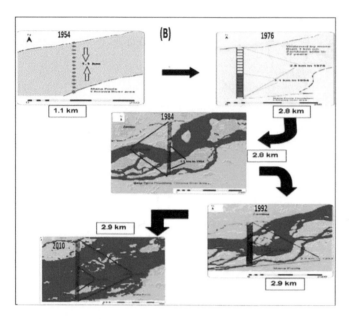

Figure 5.10: Changes in cross section of the Zambezi River near the Zambezi-Chiruwe confluence at location (B) (after Mubambe, 2012).

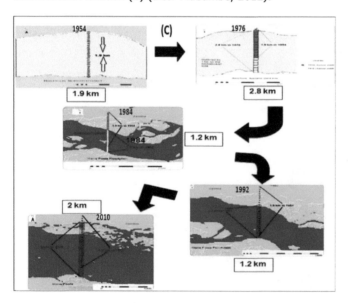

Figure 5.11: Changes in cross section of the Zambezi River channel near Nyamatusi island at location (C) (after Mubambe, 2012).

Table 5.3: Summary of the measured values of channel width at locations (A), (B) and (B) along the Zambezi River in 1954, 1976, 1984, 1992 and 2010 (Mubambe, 2012).

Location of cross section	Channel width in 1954	Channel width in 1976	Channel width in 1984	Channel width in 1992	Channel width in 2010	Total change as (%) of Pre Kariba width
(A) Zambezi River near Chongwe-Zambezi confluence	1.30	2.06	2.10	2.09	2.10	0.8 m (61.5%)
(B) Zambezi River near Chiruwe-Zambezi confluence	1.10	2.80	2.80	2.90	2.90	1.8km (138.5%)
(C) Zambezi River near Nyamatusi island	1.90	2.80	1.20	1.20	2.01	0.11km (5.8%)

These results are confirmed by Khan et al. (2014) who found that the river width doubled in the period 1963-2012.

5.3   2D Morphodynamic modelling

The main objective of using a 2D morphodynamic model is to aid the understanding of the complex morphodynamic processes of the middle Zambezi. By comparing the results from the pre Kariba and the post Kariba periods, the numerical model allows assessing the role of the dam on the observed changes. Part of the work is described in the published paper Khan et al. (2014) and in the MSc work by Khan (2013).

The model was constructed on scarce available data and was not fine tuned. Therefore the model results cannot be used to analyze any local aspects, but to assess the general trends of the river.

The model was constructed using the open-source Delft3D software (www.deltares.nl), designed for applications in the field of river and coastal engineering where water flow, transport of water-borne particles and water quality play an important role and need to be modelled to aid a better understanding of complex hydrodynamic systems. The Delft3D package is composed of several modules that can be used separately or in combination. For the Middle Zambezi, only the DELFT3D-FLOW module was used, which allowed simulations of river hydrodynamics and morphodynamics in two dimensions (2D) (Delft-Hydraulics, 2006).

DELFT3D-FLOW simulates the flow of water by solving the Navier-Stokes equations using a finite-difference method. The system of 2D equations is composed of the continuity and momentum equations of water in the horizontal directions (direction x and y). The equations are depth averaged (2D) and given in Cartesian rectangular format.

The following assumptions are made that:
- The fluid is incompressible;
- The vertical length scale is much smaller than the horizontal length scale, so the velocities on the vertical direction can be neglected. This is known as the shallow-water approximation.
- Density variations are small in relation to the density of the water. This is known as the Boussinesq assumption.

The two components of the system of equations are described as follows. The continuity equation of water is given by:

$$\frac{\partial \zeta}{\partial t} + \frac{\partial (h \cdot u)}{\partial x} + \frac{\partial (h \cdot v)}{\partial y} = S \qquad (5.1)$$

In which:

ζ	water level above some horizontal plane of reference	[m]
h	mean water depth	[m]
u v	depth averaged velocity components in x and y direction	[m/s]
S	source/sink	[m/s]

The horizontal momentum equations for x and y are given by:

$$\frac{\partial u}{\partial t} + u \frac{\partial u}{\partial x} + v \frac{\partial u}{\partial y} = -g \cdot \frac{\partial \zeta}{\partial x} - \frac{g}{C^2} \frac{u^2}{h} + HDT_x \qquad (5.2)$$

$$\frac{\partial v}{\partial t} + u \frac{\partial v}{\partial x} + v \frac{\partial v}{\partial y} = -g \cdot \frac{\partial \zeta}{\partial y} - \frac{g}{C^2} \frac{u^2}{h} + HDT_y \qquad (5.3)$$

In which:

g	gravitational acceleration	[m/s^2]
C	Chezy roughness coefficient	[m$^{1/2}$/s]
HDT	horizontal diffusion term	[m/s^2]

For the study of the changes of river morphology, the equations for water (5.1, 5.2 and 5.3) are coupled to a sediment balance equation and equations describing sediment transport. Delft3D distinguishes two modules of sediment transport: bed load and suspended load. For bed load, changes of bed level are derived from the sediment balance (Exner) equation:

$$\frac{\partial Z_b}{\partial t} + \frac{\partial S_x}{\partial x} + \frac{\partial S_y}{\partial y} = 0 \qquad (5.4)$$

In which:

$\dfrac{\partial S_x}{\partial x}$ and $\dfrac{\partial S_y}{\partial y}$ = variations of sediment transport capacity in x and y directions

$\dfrac{\partial Z_b}{\partial t}$ is temporal variation of bed level

The sediment transport capacity is computed by a user-selected formula (in the present case: Engelund-Hansen (1967), and the bed level is updated according to the case (erosion or sedimentation) (Equation 5.4). After this computation the hydrodynamic model receives a feedback, so that in the next time step the bathymetry for hydrodynamic simulation is updated.

The Engelund-Hansen (1967) formula for sediment transport is expressed in Delf3D as:

$$S = S_b + S_{s,eq} = \dfrac{0.05\alpha q^5}{\sqrt{g}C^3\Delta^2 D_{50}} \qquad (5.5)$$

In which:

S	total sediment transport	[m^3/s.m]
S_b	bed load	[m^3/s.m]
$S_{s,eq}$	suspended load	[m^3/s.m]
q	flow velocity	[m/s]
Δ	relative density of the sediments	[-]
C	Chezy coefficient	[m$^{1/2}$/s]
α	calibration factor	[-]

Note that the formula computes both bed load and suspended load simultaneously without making a distinction between them. The calibration factor was adjusted so that the transport rates are as close as possible to the observed ones.

The transport of sediment in suspension is modelled using the advection-diffusion equation, without a decay process since the transported particle is a sediment particle. The source and sink terms in the sediment balance equation correspond to erosion and sedimentation, respectively.

$$\dfrac{\partial hc}{\partial t} + \dfrac{\partial huc}{\partial x} + \dfrac{\partial hvc}{\partial y} - h \cdot \dfrac{\partial}{\partial x}\left(\varepsilon_{s,x}\dfrac{\partial c}{\partial x}\right) - h\dfrac{\partial}{\partial y}\left(\varepsilon_{s,y}\dfrac{\partial c}{\partial y}\right) = hS \qquad (5.6)$$

In which:

c	concentration of sediment in terms of mass	[kg/m^3];
$\varepsilon_{s,x}$	eddy diffusivity of sediment in x - direction	[m^2/s];
$\varepsilon_{s,y}$	eddy diffusivity of sediment in y - direction	[m^2/s];
S	source and sink terms per unit area	[kg/m^{3}s];

The time scale of the morphological changes is much larger than the scale for changes in hydrodynamics. Considering that the stability of computations is assured by using very small time-steps, to simulate the morphological changes over relatively long period of time (years to tens of years) a very long computational time is required, which may not be practical. In Delft3D-Flow a morphological scale factor can be used to reduce the duration od the computations. The morphological scale factor is applied by multiplying it to the erosion and deposition fluxes at each time step, accelerating the bed-level changes. For instance, with a morphological scale factor of 50, a hydrodynamic simulation of 1 year corresponds to a morphological simulation of 50 years.

During the first time steps the model for hydraulics is still not stable and the solutions are erroneous and thus cannot be used to compute the bed-level changes. In order to discard these first time steps a spin-up time can be defined in Delft3D-Flow such that only after this spin-up time (time required for the hydraulic model to be stabilized) the morphological computations can be started.

In Delft3D-Flow the bank erosion process is reproduced at the margin of the channel, where wet computational cells (channel) are adjacent to dry cells (dry floodplains). In Delft3D, a factor for erosion of adjacent dry cells distributes the erosion rate of a wet cell to the adjacent dry cell which then becomes wet. This results in shifting the margin of wet cells towards the floodplain, mimicking the effects of bank erosion.

In Delft3D the modeller has the option to create his own grid, where cell sizes and grid shapes can be chosen to best fit the purposes of the study. The size and shape of the cells must fulfill the condition for stability of the computations and accuracy of results.

5.3.2 Model Setup for the Middle Zambezi

5.3.2.1   Computational grid
Before the construction of the grid, the model domain was defined using the satellite images from the Mana Pools area. A 15 km segment of the Zambezi River reach was selected so that the disturbances at the boundaries of the model do not affect the area of interest for the study.
During the construction of the grid, the basic principles of discretisation were taken into account. The cell sizes, orthogonality, smoothness in cell size changes were carefully worked out so that the inaccuracies and numerical errors are avoided and the simulation time is minimized. Refer to the grid as shown in Figure 5.12.

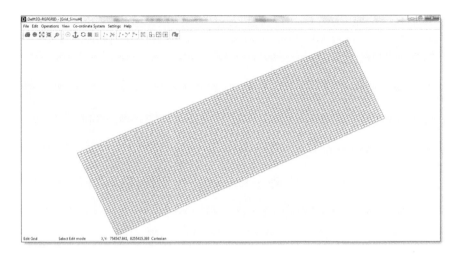

Figure 5.12: Computational grid used in the Middle Zambezi model (Khan, 2013).

Given the scope of the model exercise of studying general morphodymic trends and not local processes, the grid cell size was made to be 150 m by 100 m on average, giving an aspect ratio of 1.5. The grid covered the main channel and floodplains of the Zambezi River for 15 km, with a width of 3.5 km.

5.3.2.2   Bed topography and bed roughness

The bed elevations of the main river channel were combined with the topography of the floodplains to produce the river and floodplain bed surface to be used in the model. The elevation of the floodplain was obtained from the SRTM DEM with resolution of 90×90 m. The DEM was pre-processed in a GIS and then exported for use in Delft3D. Due to the inaccuracies of the DEM and to simplify the model, a flat floodplain was adopted with real (average) elevations. The bed elevations of the main channel were reconstructed using old satellite images. The cross-section was simplified to a rectangle with flat bed, but containing the bars observed in the satellite images (Figure 5.13).

A first estimation of the bed roughness (Chézy coefficient) was based on field data obtained from the surveys of river bathymetry and discharge measurements at Zambezi at Samango Camp (ZM4) for both low and high flows, see Figures 5.14 (low flow) and 5.15 (high flow). These measurements were carried out in different sections of the river, so using the information about cross-sectional areas, wet perimeters and discharges in the Chézy formula, the Chézy coefficient was estimated. Since there was no information available from the pre-Kariba period it was assumed that the Chézy value did not change over the years and the value corresponding to the present was adopted in all cases.

5.3.2.3   Longitudinal bed slope

The slope of the river was derived from the DEM. An analysis based on the morphological equilibrium theory (Jansen et al., 1979) was carried out to study the variations of the slope in the past (Kan et al., 2014) to establish whether the current slope could be applied also for the past conditions. The results of this analysis show that no important slope variations can be expected for the periods 1954 - 2010 (Khan et al., 2014).

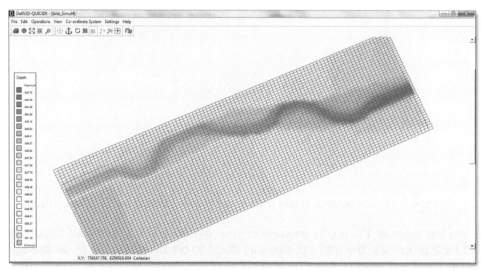

Figure5.13: Bathymetry of the middle Zambezi River reach segment (Khan, 2013).

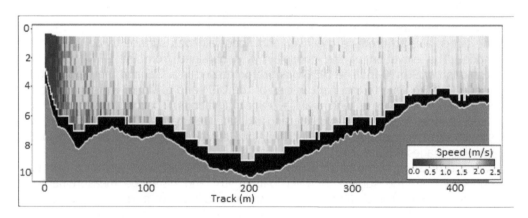

Figure 5.14: Velocity and channel bathymetry measurements (at low flows) with ADCP (field surveys of 27th November 2010 with river discharge of 1,264 m³/s).

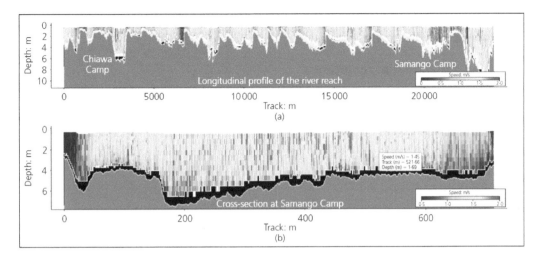

Figure 5.15: Combination of river cross section (at high flows) and longitudinal profile used as an input in reconstructing the bathymetry of the river segment in the model (field surveys of 10th March 2011 with river discharge of 4,364 m³/s).

5.3.2.4   Roughness of floodplains
The values of bed and floodplain roughness were optimized during the model calibration process to obtain water levels that are as close as possible to the measured ones.

5.3.2.5   Bed material
The characteristics of the bed sediment were derived from the analysis of the samples collected during the field campaigns (see Section 5.1.2). As it can be seen from the granulometry curves in Figure 5.6, the mean sediment size in the modeled area is 0.5 mm. The maximum size is around 1 mm and the proportion of fine sediments is low. Moreover, the mean diameter of the sediments in the main river channel is rather similar to the mean sediment diameter in the tributaries. This is probably due to the fact that most of the sediments carried by the Middle Zambezi River are brought in by these tributaries.

In the model, the representative sediment diameter was assumed to be current median diameter, assuming that this has remain constant and equal to the pre-impoundment period, since there was no information regarding the granulometry in the past. This assumption is not so accurate, but it is expected that the model is still able to reproduce the general trends in morphological development.

5.3.2.6   Boundary conditions
The upstream boundary is located near Chakanaka Farm (ZM3) and below the confluence between the Kafue and the Zambezi River. For this boundary the condition defined is a hydrograph which depends on the scenario being considered (Pre-Kariba and Post-Kariba). The amount of sediment entering the model domain from the upstream boundary was computed using the using the Engelund-Hansen (1967) formular and assuming uniform flow conditions. For the downstream boundary condition a rating curve was defined based on water level measurement

and Kariba outflows data. The sediment transport at the downstream boundary was computed by the model itself.

5.3.3 Additional parameters

5.3.3.1 Simulation time-step

The simulation time-step was first estimated using the formula below considering the maximum Courant number recommended by Delft-Hydraulics (2006), which is 10.

$$\Delta t = \frac{C_r \cdot \Delta x}{\sqrt{(g \cdot h)}} \qquad\qquad (5.7)$$

In which:

Δt	time step	[seconds]
C_r	Courant number	[-]
Δx	space step in flow direction	[m]
h	representative water depth	[m]

Delft3D uses an implicit scheme (ADI - Alternating direction implicit) which does not demand Courant numbers close to one for stability purposes. The maximum value recommended by the developer of the model was defined for accuracy purposes. After a short preliminary simulation, the time step was checked and the values were then adjusted accordingly for the definitive simulation.

5.3.3.2 Spiral motion

Since in a depth-averaged model there are no variations in z-direction, the secondary flow in river bends, also known as spiral flow, which is a 3D phenomenon, are in principle not considered. In Delft3D it is possible to take into account the secondary flow by considering additional terms in the momentum equation. To include this phenomenon the following parameters were adjusted:

- *Espir* - Parameter that enables consideration of the secondary flow in the momentum equation. Value used: 1.
- *Beta_c* - Parameter applied to the spiral flow to specify the fraction of the shear stress due to secondary flow that is accounted for in the momentum equation. Value used: 0.5.

Including the spiral flow is very important when simulating the river morphological changes because the secondary flow is responsible for the formation of 2D morphological features (bars and point bars) that are crucial in ecological studies. For instance, the point bars are usually the regeneration sites for floodplain forests.

5.3.4 Model Calibration

The calibration process was done for both the hydrodynamic component and for the morphological component of the model.

For the hydrodynamic simulation the roughness of the bed for the main channel and floodplains was first estimated based on field data and then adjusted or calibrated so that the simulated water levels are as close as possible to the measured water levels obtained from the divers installed in different sections of the river. This was a trial and error process with an initial value based on a good preliminary estimate of the bed roughness value.

Besides the water levels the distribution of the modeled velocities was observed and compared to the velocities measured in-situ (see Figures 5.14 and 5.15). The bed roughness (Chézy coefficient) was then adjusted accordingly so that the model would reproduce the same velocity distribution in the cross-section.

For the morphological component the alpha parameter of the Engelund-Hansen sediment transport formula, factors for the longitudinal and transverse bed slope effect on the sediment transport direction and the parameters for the secondary flow optimized during the calibration process.

For the sediment transport formula the parameter alpha was calibrated so that the modeled sediment transport rates are as close as possible to the measurements of the sediment transport rates that were recorded on a section further downstream on the Zambezi River for different discharges.

The longitudinal and transverse bed slope effect on sediment transport direction is taken into account by using the Bagnold (1966) formulation, where two parameters were calibrated. The calibration of these last parameters was based on the experience from past research where the optimal values for the 3D effects of the spiral flow for point bar formation were obtained. Validation of the model was not possible due to lack of another independent set of data. All the available data were used in the calibration procedure.

5.3.4 Simulation Results

The pre-Kariba simulation (B-1) was carried out to cover 45 years to study what the natural trend of the Zambezi would be without any dams. It was observed that the natural river system would tend to maintain a lower braiding degree, mostly with development of alternate bars.

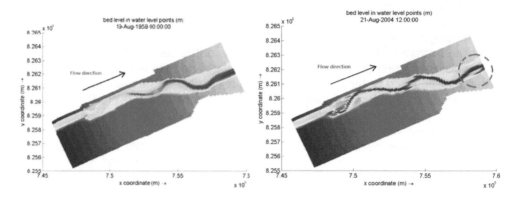

Figure 5.16: Bed levels - River planform development simulation under natural flow conditions (B-1) (Khan, 2013).

However, with altered flow conditions corresponding to the first dam operation period (Post-Kariba 1) followed by the second dam operation period (Post-Kariba 2), the morphodynamic model tends to underestimate the river width and the other morphological processes. The results of the model under natural flow conditions (B-1) and under altered flow conditions (B-2) show similar morphological developments. The average depths, Thalweg depths, channel widths and braiding degree for both simulations, B-1 and B-2 show no significant changes. The results are presented for the year 1980 (after Post-Kariba 1) and 2012 (after Post-Kariba 2), so that the effect of each of these dam operation periods can be detected. There are no significant differences between the results of the two simulations in terms of reach-averaged depths and Thalweg depths. The width variation along the longitudinal axis of the river channel results in similar river braiding degrees. The comparison between B-1 and B-2 shows no significant differences between the two cases. Refer to Figure 5.17.

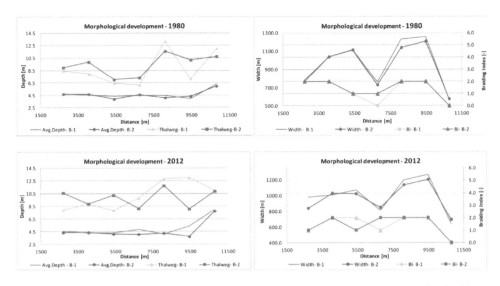

Figure 5.17: Comparison of simulation results for Pre Kariba (B-1) and Post Kariba (B-2) in terms of river width and depth showing no significant difference (Khan et al., 2014).

5.3.5 Interpretations of model results

The morphodynamic model was not able to simulate the observed morphological changes, and particularly the observed river channel widening described in Sections 5.2.1 and 5.2.2 (historical map and satellite imagery analysis). This is most probably due to the simplification in the bank erosion formulation, which was not able to reproduce the phenomenon of bank retreat in a realistic way. The model distributes the erosion occurring in the cells near the channel margin to the adjacent dry cells, which then become a part of the main river channel. This approach imposes a reduction of bed erosion in the originally wet cells (since a percentage is assigned to the adjacent dry cell), which results in a shift of the water flow towards the centre of the channel in the next computational steps, hereby reducing the near-bank flow. The higher the percentage of erosion that is transferred to the dry cell is, the stronger is the effect of shifting the flow towards the centre of the channel and the less is bank erosion. This could be seen as a rough representation of the effects of local deposition of the bank material after failure, which remains near the channel margin for a while until the flow has washed it away ("slump block armouring" in Parker et al., 2011). However, in real rivers the fallen bank material settles at the foot of the bank, whereas in the model it is distributed in one entire computational cell having an average width of 100 m. The new wet cell is also 100 m wide. This means that the effects of the bank are felt in the model for a distance of 200 m, whereas in real rivers these effects are felt in only within a few metres from the bank. The result is that the model is not able to reproduce the near-bank flow, which means that the model cannot correctly reproduce near-bank erosion and channel margin shift. Another important model shortcoming was the not taking into account the most important factor leading to bank retreat in the study area, which is the seepage force caused by groundwater movement towards the river after a sudden drawdown in the river water levels (Rinaldi and Casagli, 1999; Rinaldi et al., 2004 and 2008), as analyzed in Section 4.3.3 of this thesis.

Due to the failure to reproduce the width changes, the results of the morphological model show no significant changes in planform or braiding degree between the natural and the altered flow conditions, failing to reproduce the observed effects of the Kariba Dam. The observed short coming of morphodynamic modelling exercise supports the conclusion that the bank erosion phenomenon is the major process triggering most of the observed changes in the Middle Zambezi River morphology. In case the Kariba Dam continues to be operated such that the discharge drop suddenly and out-of-season floods are spilled, more bank and bar erosion may occur which will not allow the river to re-establish it original planform with the narrower channel.

5.4 Conclusions

It is apparent that the hydropower infrastructure and associated operations have had an impact on the morphodynamic processes of the Middle Zambezi River reach. From the analysis, it is important to note that the local catchment area and its tributaries have continued to contribute significantly to the supply and distribution of sediments, which is key to the morphodynamic processes of this river reach. However, an important phenomenon of artificially-induced channel widening has been observed in this floodplain. The causes which trigger this phenomenon being the manner in which the dam flood gates are operated is covered in greater detail in Chapter 4. This chapter observes that the effects are quite significant.

CHAPTER 6: FLOOD PLAIN VEGETATION

This chapter deals with the interaction of floodplain vegetation with the hydrodynamic and morphodynamic processes. A clear understanding of this interaction is important in the definition of possible mitigation measures that can be implemented in order to sustain the floodplain ecosystem. Hughes et al. (2012) argues that Floodplain forests are at the receiving end of all the physical, biological and human activities taking place in a catchment and therefore reflect the patterns of delivery of water, sediment and nutrients through the flood events to the floodplain. Mc Clain (2013) puts an emphasis on the need to balance development with environmental sustainability through understanding of the boundary conditions imposed on water and the environmental water requirements of affected ecosystems. Due to the riparian distinctive life requirements, in particular aspects that relate to seed and seedling development, it may be predicted that they would be more vulnerable to impacts from river damming and flow regulation (Rood et al 2010)

The work in this chapter has benefited from the MSc research by Ncube (2011), Mubambe (2012), Gope (2012) also published in Gope et al. (2015) and Khan (2013).

6.1   Choice of biological indicator

The Middle Zambezi can be considered as a unity ecosystem, providing services and benefits. Field work ("boots on the ground") is important to gain an understanding of the linkages and system interactions. The understanding of the linkages would lead to the identification of the ecosystem components that may be under strain, leading to the choice of a credible biological indicator. For the Middle Zambezi floodplains, the identification process of the biological indicator started during the Power-to-flow program reconnaissance survey of the Mana Pools area in February 2010. The field team had an impression that there was something happening to the riparian floodplain tree stand, which was characterised by very old trees, most of them dying off with a complete absence of younger trees. This gave an indication that this component (*Faidebia albida* tree) of the ecosystem was under strain and was under high risk of being wiped out from the floodplains (Figure 6.1 shows the tree stand with a lot of die-offs). This finding was in agreement with the assertions made by researchers that the spectacular *F. albida* tree was showing signs of degradation from as far back as 1970s and 1980s (Attwell, 1970, Cumming, 1983). Before making the choice, research was done on the nature and relevance of this particular tree to the floodplain habitat. Research work on the riparian *F. albida* tree was done through field surveys and in controlled environmental experiments in the framework of this research.

Figure 6.1: Mana Pools floodplain riparian tree stand showing older *Faidherbia albida* trees with a lot of tree die-offs.

6.1.1 Characteristic of the *Faidherbia albida* tree

The *Faidherbia albida*, commonly called apple-ring, is a deciduous legume tree, growing up to 30 m high with a deep taproot, going down to 40 m. The branches bear paired thorns and its leaves are pinnate with 6-23 pairs of small oblong leaflets. Flowers are arranged in yellow spikes, fruits (pods) are twisted and shiny orange, 5 cm wide and may go up to 25 cm long (Orwa et al., 2009), Figure 6.2 shows the fruit. *Faidherbia albida* has an inverse phenology by shedding their leaves during the wet season, while the leaves mature and the pods ripen during the dry season. It is also tolerant to water logging and salinity (Orwa et al., 2009).

The leaves, bark, pods and seeds of the browsed shrubs and trees constitute a vital portion of the ruminant diet providing essential vitamins, minerals, and protein which cannot be obtained from grass forage alone and without which grass forage cannot be fully utilized (Wickens, 1980). The intake of foliage is especially critical during the dry season. *F. albida* often provides the only available "green" intake during the dry season. Wentling (1983) proposes that it is evident that this tree's inverse growing cycle makes it "just what the doctor ordered".

The *F. albida* is known to have a fast rate of growing taproot, being 2 - 5 meters/year and this grows straight down for over a meter before branching out, reaching up to 30 to 40 meters. Therefore, it does not compete for surface soil nutrients and water. It has also been calculated that the annual leaf fall from 50 trees per hectare is equivalent to applying 50 tones of manure in an hectare (LeHouerou, 1980b).

Due to the value this tree provides to the environment, it was listed by the FAO Panel of Experts on Forest Gene Resources, as one of the priority species for the improvement of environment (FAO 1977, Palmberg 1981). The value of this tree provides the much needed nutrients to the animals, as outlined in Table 6.1.

Figure 6.2: Fruit of *Faidherbia albida* tree which is well sought after by animals for food.

Figure 6.3: *Faidherbia albida* tree stand at Maudas Crocodile Farm Camp showing an inverse phenology, (in leaf during dry season (September 2012) and sheds its leaves during the rainy season (January 2013).

Table 6.1: Nutritional value of the fruit leaves and tree Bark of F. *albida* (adapted from Heuzé and Tran, 2013).

Main analysis	Unit	Average	Min	Max
Dry mattet	% as fed	38.5	30.7	44.4
Crude protein	% DM	15.1	11.6	25.1
Crude fibre	% DM	20.4	12.4	33.0
Gross energy	MJ/kg DM	18.3		
Minerals	**Unit**	**Average**	**Min**	**Max**
Calcium	g/kg DM	17.0	10.0	33.6
Phosphorus	g/kg DM	1.8	0.9	2.5
Potassium	g/kg DM	10.6	5.2	15.9
Sodium	g/kg DM	0.4	0.0	0.9
Magnesium	g/kg DM	3.5	2.4	4.9
Manganese	mg/kg DM	63	53	73
Zinc	mg/kg DM	27	26	29
Copper	mg/kg DM	9	6	13
Iron	mg/kg DM	1171		

(DM = Dry Matter)

6.1.2 Distribution of the *Faidherbia albida* tree

According to Wickens (1969) and CTFT (1988) the *Faidherbia albida* tree can be found throughout Sub-Saharan Africa in the Sahel and Sudan zones of West Africa and the Eastern and Southern Africa. The tree is strongly associated with alluvial soils along perennial or seasonal watercourses. It occurs in areas with 500 to 800 mm annual rainfall. It can also be found in the northern Sahel and Sahara, in areas with good groundwater table. Figure 6.4 shows the general distribution of the tree in Africa. The tree is also commonly found in the Middle-East and in South-East Asia, India, Pakistan, Cyprus, Cape Verde and Peru (Orwa et al., 2009).

In the Middle Zambezi floodplains, *F. albida* tree species form the riparian woodland. It is therefore a special characteristic tree of the Middle Zambezi floodplains and contributes to the quality of the habitat. The tree is found on the main floodplains, on islands and sandbars (Figure 6.5 shows young *F. albida* on recently deposited sand bar and Figure 6.6 shows a mixed tree stand of *F. albida* on an island).

Figure 6.4: Distribution of *Faidherbia albida* tree in Africa (adapted from Wickens, 1969).

Figure 6.5: *Faidherbia albida* shrub getting established on recently deposited sand bar in Middle Zambezi (September 2013).

Figure 6.6: A mixed tree stand of *Faidherbia albida* on Elephant Bone Island.

6.1.3 Conditions for regeneration of the *Faidherbia albida* tree

Many trees and shrubs found in floodplain forests have quite specific requirements for their regeneration (Hughes et al., 2012). Hughes (1990) observed that for the Tana floodplain forest, low regeneration levels reflected its dependence on periodic favorable hydrological conditions. Hughes (1994) further stresses that the exact regeneration needs and tolerances of floodplain species vary hugely, not only among parts of the floodplain but also between floodplains in different geographical regions. It is important to understand the regeneration needs and mechanisms of floodplain species in order to understand the role that a particular disturbance has played in their distribution (Hughes, 1994). Hughes (1997) observes that where floodplain vegetation vitality is concerned, flood regimes do greatly influence the availability of areas for vegetation regeneration from year to year and determines the age structure of floodplain communities. In the Tana floodplain forest, Hughes (1988) identified a number of factors contributing to the lack of regeneration which included elephant activities and browsing by other animals. Hughes also identified the infrequence of suitable conditions for germination and seedling establishment; shrubs seemed to have been over browsed and therefore concluded that there is a high possibility that conditions for successful establishment of environmentally sensitive species only occur in some years.

There are three key conditions for the regeneration of the *F. albida* tree, being initial soil moisture, absence of browsing pressure and availability of groundwater. Figure 6.7 shows a schematic outline of these conditions.

6.1.3.1 *Soil moisture to aid seed germination and seedling establishment*
Soil moisture is required for the germination of the seed. The seed can germinate with or without passing through the digestive system of animals, although it makes it easier for the germination process when the seed has passed through the animals' gut. Soil moisture is also required for the establishment of *F. albida* seedlings, as they require to grow roots fast enough to tap into the groundwater source before surface layers dry out. The seedling tap root can grow 1 cm a day until it reaches a perennial water zone and up to 5 m in a year (Vandenbeldt, 1992). Failure to do this can cause seedling death. For the Middle Zambezi, the soil moisture that is needed in the initial growth stage is provided by the local catchment area rainfall.

6.1.3.2 *Absence of browsing pressure in early stages of growth*
Although *F. albida* seed is adapted for dispersal by herbivores, being hard and tolerant of the digestive process (Timberlake, et al., 1999), its establishment is believed to be greatly affected by browsing pressure. In the Tana floodplain forest, Hughes (1988) and (1990) also identified grazing and browsing by wildlife as one of the factors of importance in determining floodplain vegetation distribution. It has been observed that normally very few seedlings survive the first year, as they are browsed by herbivores or eaten by insects. Major browsing impacts are observed when browsing happens before the tree produces its protective spine, which it does in a few months (Barnes and Fagg, 2003). Therefore, to assure floodplain forest rejuvenation, there should be absence of browsing pressure until the seedlings are well established (Barnes and Fagg, 2003). Floodplain forests are flood-dependent ecosystems, often linked to well-timed, periodic floods for the provision of regeneration sites. Further, regeneration flows are often synonymous with flood flows and only occur periodically (Hughes and Rood, 2003).

6.1.3.3   Availability of stable groundwater source

The dry season growth of *F.albida* relies on uptake of water from deep subsurface or watertable making it a facultative phreatophyte (Roupsard et al., 1999), in that it habitually obtains water supply from the saturated subsurface zone. Dunham (1991) confirmed the independency of the tree's water supply on surface water, after noticing that there was no influence on the phenology of *F.albida* on Mana Pools, despite the occurrence of the greatest flood in 1981. The tree has the ability to develop a deep tap root that can grow to about 30 to 40 metres, enabling it to tap into deep groundwater resources (Barnes and Fagg, 2003). Due to this ability, *F.albida* was reported to be less affected by severe drought stress (Roupsard, et al., 1999), which was attributed to the tree's rooting system distributed among three main soil compartments. These are; (1) abundant fine roots in the superficial layers; (2) the majority of roots between 2 and 4m depth; (3) and appreciable number of roots down to the zone of capillary rise from the water-table (7.5 m). However, despite the abundance of fine roots in the superficial layers, Roupsard et al. (1999) reported low contribution of these roots to the tree's annual water uptake. These roots were, however, implied to remain hydrated in the dry season to enable the plants to make use of early rain events.

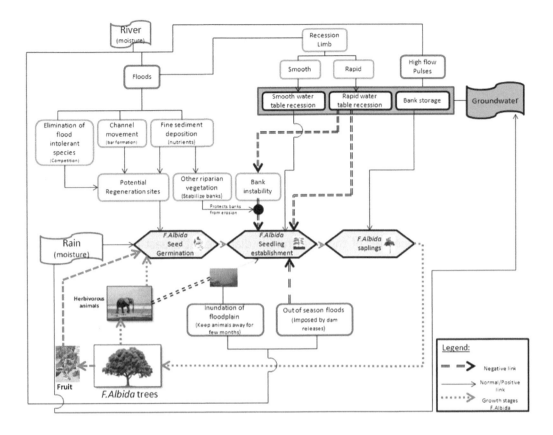

Figure 6.7: Schematic representation of the system for survival and regeneration for F. albida (adapted from Khan, 2013).

6.2 Field surveys

6.2.1 Transects and tree sampling plots

Systematic sampling was carried out in the *F. albida* dominated woodland on the Mana Pools floodplains of the Middle Zambezi. In the first field campaign, a total of 23 plots were surveyed on eight transects on the main Mana Pools floodplains (Figure 6.8) and in the second field campaign transects were established perpendicular to the river from the islands stretching out to Mana floodplains (Figure 6.9).

In the first field campaign, the transects were spaced 1 km apart, except where there were obstacles like streams that could not be crossed. The 50 m x 50 m vegetation plots were set out to charter for the pixel size (30 m x 30 m) of the satellite images used in section 6.3, to analyse the changes on the Mana Pools floodplains. A tape measure was used to mark the plot boundaries, while coordinates of the centre of the plots and spacing of the transects was measured using the Garmin GPSmap 60CSx Geographical Positioning System (GPS) (Figure 6.10). This approach of using vegetation plots was adapted from similar forest and vegetation based studies done by other researchers, such as Carter et al. (1995), Maingi and Marsh (2002), McLaren and McDonald (2003), Hitimana et al. (2004), Nebel and Meilby (2005), Thoms et al. (2005), Mwavu and Witkowski (2009) and Powers et al. (2009).

The transects were laid out perpendicular to the river across the floodplain on straight lines, marked using a GPS. Slight diversions were done where there were barriers like over flowing streams, since the field work was carried out at the start of the rain season (December 2010 and February 2011).

In the second survey, shown in Figure 6.9, since islands were small and had a narrow width, only one plot was established on the islands, whereas two were set on the floodplain. However, transect number four had two island plots, because it had two adjacent islands both with *F.albida* trees. The island with transects 7 and 8 was also fairly long thus two transects were established on it (Figure 6.9).

Each tree plot was characterised by GPS coordinates, number of trees, height of each tree and diameter of the tree trunk - diameter at breast height (DBH). Figure 6.11 shows the tree characterisation in each tree sample plot.

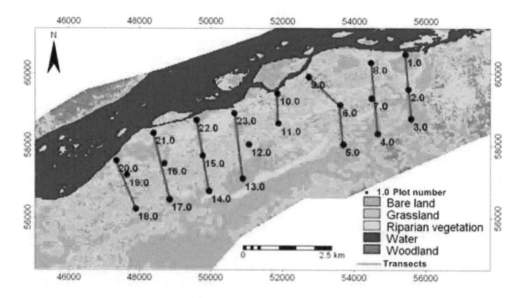

Figure 6.8: Transects and 23 sample plots for characterisation of the *Faidhebia albida* stand on the Mana Pools floodplains (Ncube, 2011).

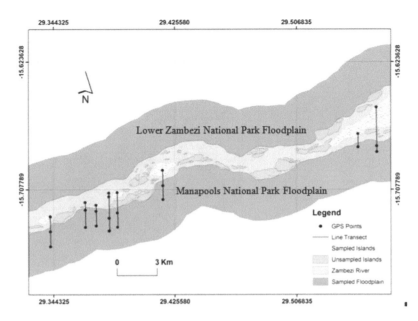

Figure 6.9: Second field campaign with 8 transects and 24 sample plots for the combination survey of the floodplain and the islands (Gope, 2012 and Gope et. al., 2015).

Figure 6.10: Setting out of tree sample plot using a tape measure and GPS during the February 2011 field survey.

Figure 6.11: Tree characterisation in each plot - (number of trees, DBH and heights) during the February 2011 field survey.

The analysis of the survey results shows a fully grown tree stand on the floodplains, with DBH ranging between 41 and 200 cm, whereas the islands plots showed a younger and well mixed tree stand. This indicated that the trees were regenerating only on the islands and there was no regeneration on the main floodplains (Figures 6.12 and 6.13). The results also show a corresponding difference in the height distribution for the floodplain trees and the island trees (Figures 6.14 and 6.15)

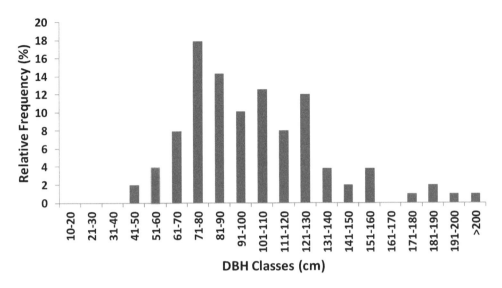

Figure 6.12: Tree stand characterisation in terms of dbh for the floodplain plots (*n*=94), showing an older tree stand with no young trees (Ncube, 2011, Gope, 2012, Gope et. al., 2015).

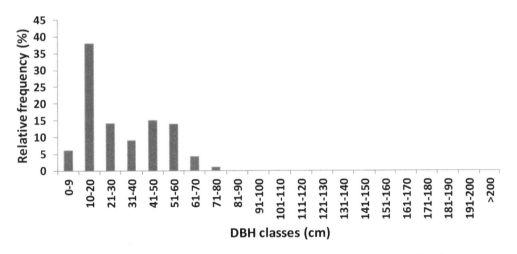

Figure 6.13: Tree stand characterisation in terms of dbh for the island plots (*n*=38), showing a younger and healthier tree stand (Gope, 2012 and Gope et. al., 2015).

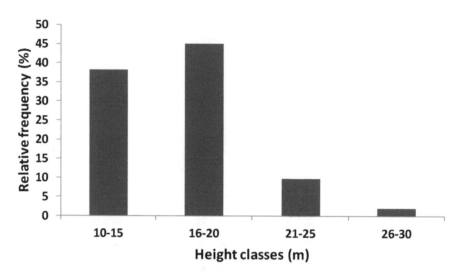

Figure 6.14: Tree stand characterisation in terms of height for the floodplain plots (*n*=94), showing an older tree stand with no young trees (Ncube, 2011, Gope, 2012 and Gope et. al., 2015).

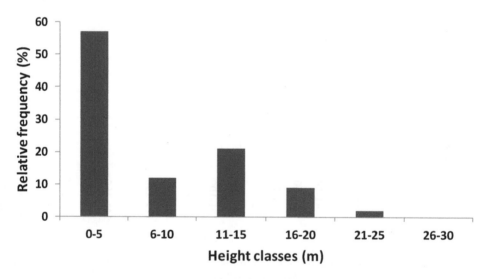

Figure 6.15: Tree stand characterisation in terms of height for the island plots (*n*=38), showing a younger and healthier tree stand (Gope, 2012 and Gope et. al., 2015).

6.2.1 Extraction of a tree ring cores and tree disks

In an effort to date the forest stand, two methods were used, these were the extraction of tree ring cores using the incremental borer and slicing of trees to get the disks for tree rings.

The method of extracting the core entails the use of an incremental borer to extract the tree cores which are then staged on a microscope to count the number of rings with each ring corresponding to the year that the tree has grown thereby allowing for the analysis of the growth trends of a tree. An incremental borer manufactured by Haglof with a diameter of 12mm and length of 300 mm was purchased to extract the cores of sampled *F.albida* trees on the Mana floodplains. Two to three cores were extracted per one vegetation sample plot. The cores were extracted from the dbh height (1.3 m) of a tree (Worbes et al., 2003). The incremental borer was penetrated into the tree trunk while being rotated to allow for the slicing of the tree cores. This was done until the threads of the incremental borer were engaged and the tip of the auger was embedded in the tree trunk. Thereafter, the core extractor was inserted in the auger to remove the core (Grissino-Mayer, 2003). Figure 6.16 shows the process that was undertaken when extracting the tree cores. The extracted cores were then put in plastic papers for analysis. However, the cores extracted as shown in Figure 6.16 were not analysed because it was found that the cores that were extracted were too short due to the size of the borer used.

The dating of the trees was then only based on a tree disks (Figures 6.16 and 6.17). Although only four tree disks were collected, this gave an indication on the correlation of the size of tree in terms of (dbh) and the age. The counting of *F.albida* tree rings was the method that was adopted in order to age the current *F. albida* stand structure on the Mana floodplains. Research has established that the rings of this tree species are indicated by the narrow bands of marginal parenchyma which mark the annual growth periods and hence can be used for age determination (Barnes and Fagg, 2003). Ring widths are said to be correlated with the rainfall recorded in each wet season and the annual minimum temperature for this species.

Figure 6.16: Extraction of a tree ring core process during the February 2011 field survey.

According to Gourlay (1995) *F. albida* is generally problematic in ring detection, therefore tree cross-sections presented less challenges and were preferred to the use of cores. Stem cross-section disks were collected from four fallen *F. albida* trees since the study area is within a protected National Park. However, the sampled fallen trees were still alive as some branches were still active. Two trees were cut from the islands and the other two from Mana Pools floodplain. Trees were cut with a chain saw and two stem disks of about 5 - 7 cm thick were collected from each tree. The disks were cut at about 2 meters height from the ground.

The disks were air dried for about 6 weeks and slices were extracted from 3 samples from the floodplain. Two slices were cut from each disk and was brought to Netherlands for analysis. The cutting of slices was done to reduce weight during transportation from the study country to Netherlands. Slices were cut from the pith in nearly opposite directions except in situations where the other side of the disk had damages like scars (Figures 6.17 and 6.18). Areas with smallest rings or widest rings were avoided as these areas usually consist of reaction wood.

The results show that *F. albida* trees formed wide trees rings close to the pith indicating fast growth of the tree in its juvenile phase, the phase after establishment. This holds especially true for the two trees sampled on the islands, where in tree ZWIS3MC, tree rings of up to 2 cm width were formed in several years (Figure 6.19). In the older trees a decline in ring width with increasing age was observed, tree ZWFL2 showed quite a strong decrease in radial growth after about 20 years (Fig 6.15) and at radius of about 10cm. This is comparable to the slowing down of incremental growth of the same species at diameters of about 10-15 cm stated by Wood (1989). The formation of wide rings in the early growth phases of the tree made ring detection relatively easy in young trees and in the early phases of old trees. The observations on the growth rate of the tree in early phase confirm the results of other researchers where juvenile trees were stated to grow very fast (Barnes and Fagg, 2003). However, as the tree gets older ring widths become narrow (Fig 6.20). Older tree ZWFL1 showed an average ring width of about 1.4 mm during the last 30 years.

Examination of the cumulative radial increment data shows several important patterns. Island tree, (number 4) reached 10 cm radius at 11years of age, while floodplain trees (number 1 & 2) reached the same radius size after 20 and 24 years, respectively. There was some variation in the growth rate of island trees (tree 3 and 4) after the first 6 years. For the floodplain trees, a 50 cm diameter corresponds to trees of about 100 years of age (96 years for tree number 1 and 79 years for tree number 2) (Gope, 2014).

Figure 6.17: An illustration on how the analyzed slices from cross-section disks were cut to reduce weight during transportation (Gope, 2012).

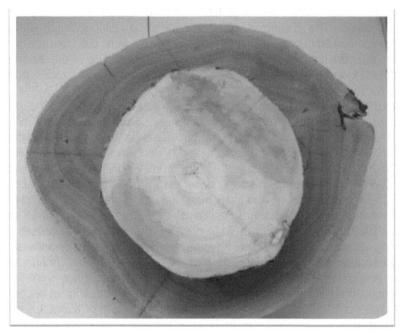

Figure 6.18: Showing the differences in colour of wood between a young *F,albida* tree (small disk) and an old tree (large disk) (Gope, 2012).

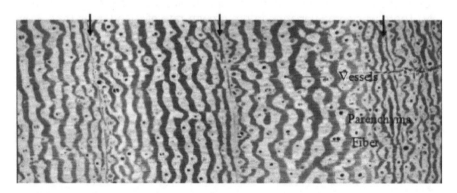

Figure 6.19: *F.albida* macroscopical, black arrows indicate annual ring boundaries and blue arrows indicate different cell types (Gope, 2012, Gope et al., 2015).

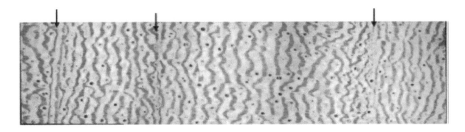

Figure 6.120: Young *F.albida* macroscopical, black arrows indicate ring boundaries (Gope, 2012).

6.3   Laboratory investigations

This activity was carried out in the field at three independent plots and in a backyard to study the following: 1) assess effectiveness of *F. albida* seed germination from elephant dung; 2) density of seedlings in the field at the commencement of the rain season; 3) germination and growth in absence of animal browsing, in a controlled environment and; 4) observe how and whether the *F. albida* seed which had not been eaten by animals would germinate.

6.3.1 Field plots observation
Two field observation plots were established at the Chiawa camp and at the Chongwe river camp. At the Chiawa camp field plot, observation was made on the effectiveness of the *F. albida* seed germination from the elephant dung (Figure 6.21). At the Chongwe River camp field plot, observation was made on the general density of *F. albida* seedlings during the early part of the rain season (Figure 6.22).
The results show that the *F. albida* seed does germinate in large numbers with a high density per square meter, but due to browsing pressure all the seedlings are eaten by the animals before they get established. By the end of the rain season, no seedlings could be seen in the two field observation plots, confirming that the browsing pressure is too high on the floodplains leading to negative consequences for regeneration of the *F. albida*.

Figure 6.21: *F. albida* seed sprout much easier when passed through the animal digestive system. Shows the sprouting of seed from the elephant dung at Chiawa Camp field observation plot.

Figure 6.22: Location of the 1 m^2 observation plot with over 30 seedlings of Faidherbia albida - seedling germination at the Chongwe River camp field observation plot (January 2013).

6.3.2   Planted trees in agro forestry at GART Research Centre

Two sample observation plots were established in a controlled environment. The Chisamba GART Research Station observation plot was to observe how the *F. albida* grows in areas where there is no browsing pressure. While the backyard garden observation plot was to observe how and whether the *F. albida* seed which had not been eaten by animals would germinate.

At Chisamba GART Research Station, the observation plot comprised two agro forestry fields of 5 year old (Figure 6.23) and 15 year old (Figure 6.24) planted trees. In the absence of younger trees in the study area, the observation was to determine how the *F. albida* tree grows in an area without animal browsing. The tree ring cores extracted were to be used as control cores for the determination of the rings for dating purposes. This controlled environment was requested by the forestry laboratory technician who was going to analyse the tree ring cores which were extracted in Mana Pools floodplain sample plots.

The results showed good growth pattern, with clear differences in terms of dbh and hieght between the younger 5 year (with dbh - 15 cm, hieght - 8 m) old tree and the older 15 year (with dbh - 25 cm, hieght - 12 m) old tree. The Research station confirmed they had no problem propagating the trees which were used in agro forestry demonstration plots.

Figure 6.23: Charaterisation process (hieght, DBH) and core extraction - 5 year old *F. albida* tree at Chisamba GART Research.

Figure 6.24: Tree characterisation of a 15 year old F. albida tree - height, DBH and extraction of a core.

6.3.3   backyard garden

The main objective of the backyard investigation was to find out how the seed germinates and to examine the early stages of tree growth. To do this, seeds were picked and placed in a tab with soil in the researcher's backyard garden. After watering for one month at the frequency of four times in a week, the seeds started to sprout. The sprouting process went on for another two months with continuous watering (Figure 6.25).

In the backyard, the seedlings easily got established, such that after 6 months it was difficult to transplant them, as the roots had become well established. This showed that even without passing through the animals' digestive system, the seeds would still be propagated in nurseries and at the end of six months, the trees' root system would be well established.

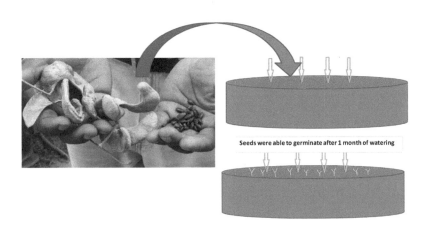

Figure 6.25: Back yard garden observation - *F. albida* seed was able to geminate after one month of watering, showing good established root system after 6 months.

6.4   Satellite image analysis

Though riparian forest degradation has been observed by a number of researchers, the most recent analysis using satellite image was undertaken by Mubambe (2011). A clear evidence of the lack of regeneration was presented from the analysis by observing the trends from images of 1954, 1984, 1992 and 2010 (Figure 6.26). The 1954 image shows a very high density of forest stand, presenting almost no bare land. However, already in the 1984 image, bare land is present. The surface covered by bare land further increased with time, and by 2010 the forest stand is limited to strips interspersed by bare land. The images also show that the sand bars and islands that had no trees in 1954 are colonised by trees by the year 2010. The image of 2010 shows also that there has been a significant die-off of old trees on the floodplain with little or no regeneration after 1954.

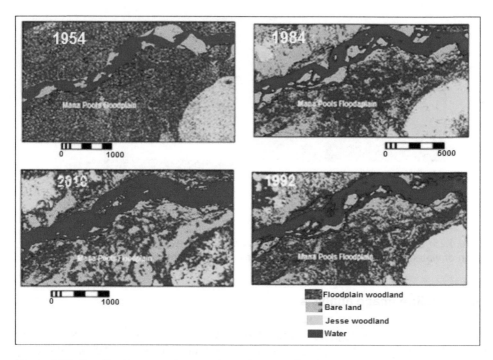

Figure 6.26: Satellite imagery showing degradation of floodplain riparian woodlands over the years - 1954, 1984, 1992 and 2010.

6.4   Conclusions

The tree characterization indicates that there are some difficulties in regeneration on the floodplains. However, the tree stand on the islands and sand bars shows a healthy mix of tree ages, including young trees, with the height distribution (55%) clustering in the class of 0 - 5 m and 5 - 15 m and few old trees with height of more than 20 m. The *F. albida* tree has started colonising the previously bare islands and sand bars. Further, the stands on the islands show that the trees are not under pressure from the animal browsers. This research shows that floodplain trees are older than 80 years, which indicates that the tree stand was established before the hydropower schemes. The absence of young trees shows that with the current dam operations, the floodplain will only maintain the older trees and once they all die, that would be the end of this riparian tree on the floodplain. The field observation and the controlled observation plots show that trees need initial soil moisture to germinate and get established. Without browsing pressure, the tree is able to establish its root system within a period of six months and within one year the tree is able to tap into ground water resources. The need to cut off the browsing pressure during the early stages of growth holds the key to the trees' regeneration. This research then observes that the permanent residence of browsers on the floodplain will not allow the growth of this important tree.

CHAPTER 7: SUGGESTED ENVIRONMENTAL FLOW REGIME

This chapter seeks to provide the framework and the justification for the recommendations related to the establishment of an environmental flow regime for the Middle Zambezi River reach. As described in Sub-chapter 1 of this thesis, the Middle Zambezi consitutes an important floodplain ecosystem, providing a rich habitat for high biodiversity of wildlife. In this chapter therefore, the Middle Zambezi River reach will be considered as an environmental ecosystem whose need for water and other interventions is highlighted.

The definition for an environmental flow regime given in this chapter relates closely to the definition of Krchnak et al. (2009). Environmental flows (EF) consist in a variable water flow regime designed and implemented as intentional releases of water from a dam into a downstream reach of a river, in an effort to support desired ecological conditions and ecosystem services. Environmental flows are necessary to maintain the health and biodiversity of downstream water bodies and their floodplains. Besides the amounts, one should also specify when these flows should be ensured. This departs slightly from the common understanding of EFs being "the quality, quantity and distribution of water required to maintain the components, functions and processes of a riverine ecosystem on which people depend" (O'Keeffe Jay and Tom Le Quesne, 2009, Acreman and Ferguson, 2010). Rood et. al.(2005) argue that rather than seeking to restore a river system to its pre-development condition, a more practical objective might be to establish a smaller (or larger) river system that displays the same essential ecosystem functions as the original river, but has been scaled to reflect the new hydrologic situation. As it is unlikely that pristine, pre-development riverine conditions can ever be recovered, therefore, restoring critical ecosystem functions may be a more feasible objective.

The case of the successful restoration of the lower Truckee River and the achievement attained by modifying the patterns of flow regulation provides optimism for the conservation and restoration of other dammed rivers, that restoration may be achieved without sacrificing other water commitments, since the restoration efforts are particularly targeted toward high-flow years (Rood et. al., 2003).

The work presented in this chapter is based on the synthesis and combination of all research results covering hydrodynamic, morphodynamic and vegetation interactions that exist in the hydropower-dominated study area. The main question to be answered is: in view of the observed processes and interrelations, can the dam operating rules be optimised in order to save the downstream floodplain environment from complete degradation. The area of emphasis is the morphodynamic process response that has led to channel widening and the lack of regeneration for the floodplain riparian tree, *Faidherbia albida*. The work acknowledges that some changes that have already occurred are not reversible, but some aspects that can be mitigated have the potential of maintaining the health and sustenance of this important ecosystem.

7.1 Literature review

Environmental flow regime is not the minimum flow that has to be allocated to a river, it is a series of discharges, including low and high water flows, minimising the impact of flow regulation. When the flow discharge in a river is changed, the entire system is altered (Crosato, 2014; Lu and Siew, 2006; Richter and Thomas, 2007; Bunn and Arthington, 2002; Matos et al., 2010). Richter and Thomas (2007) argue that the alteration of natural water flow regimes brought by dam construction and operation has had the most pervasive and damaging effects on river systems.

Riverine ecosystems include abiotic (physical environment made of water and sediments) and biotic (flora and fauna - both aquatic and terrestrial) components. These components interact and influence each other at all spatial and temporal scales. In this regard, it is not possible to define a sustainable management solution without considering and understanding the interactions between the different components. It is this understanding that allows identifying the ecosystem components that may be under strain due to impacts of flow regime alterations and how restoration and sustainability of the system can be achieved. Failure to understand these interactions would lead to environmental flow prescriptions that do not foster sustainability (Department of Water and Sanitation- DWS, 2014). It is important to note that once a river reach has been disturbed, it may never be possible to return to the pristine conditions before the disturbance, therefore, only what is feasible and what would yield the required results would be supported (Jain, 2012).

O'Keeffe and Le Quesne (2009) identified over 200 methods for assessing environmental flows, which can be divided into the five broad categories of: lookup table approaches; extrapolation approaches; hydraulic rating methodologies; habitat simulation methodologies; and holistic methodologies. The choice of the assessment the category is said to depend on urgency of the problem, resources available for the analysis, importance of the river, difficulty of implementation, complexity of the system.

Although O'Keeffe and Le Quesne (2009) argue that lack of information and resources should never be a barrier to some implementation of environmental flows and that fine-tuning would be done as more knowledge and resources become available, Hughes (2001) and Department of Water and Sanitation (2014), emphasise the need to have a good and sound data and information base on which decision making can be based for successful definition and implementation of environmental flows.

The concept of environmental flows was introduced in southern Africa in the late 1980s, and was subsequently included in the national legislation of several southern African Development Community countries. While much of the focus in the intervening years has been on the assessment of EFs for new water resource developments, some attention has been paid to the possibilities of modifying the operation of existing infrastructure to deliver EFs (Brown and King, 2011).

7.2   Description of the suggested environmental flow regime

To save the floodplain riparian tree *Faidherbia albida* and to reduce channel widening, the environmental flow regime of the Middle Zambezi should have two components, operating at two different temporal scales.

7.2.1 Saving the floodplain riparian tree *Faidherbia albida*

From the analysis in Chapter 6, the browsing pressure from the animals resulted as the main factor for the lack of regeneration of the *Faidherbia albida* tree. In the pre-Kariba era the occasional flooding used to keep the animals away from the floodplain for a period of time of months. Now the animals are resident on the floodplain throughout the entire year (Figure 7.1, showing the difference in the flooding frequency between the unregulated and the regulated flows). Therefore, a deliberate flood is recommended once every five years to keep away the animals and allow for tree regeneration.

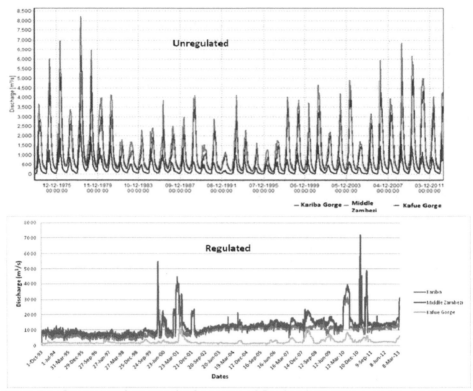

Figure 7.1: Comparison between the frequency of flooding between unregulated and regulated situation of the Middle Zambezi.

7.2.1.1 Nature of flood to keep the animals from the floodplain to facilitate tree regeneration

From the analysis of the pre-Kariba period, the unregulated system would present the following floods:

- (>8,000 m^3/s) - extreme high floods;
- (6,000 to 7,000 m^3/s) high floods;
- (4,000 to 6,000 m^3/s) moderate floods; and
- (3,000 to 4,000 m^3/s) low floods.

From the past (mostly oral record), the area is known to have experienced also some exceptional floods which covered the area extensively, as can be seen in Figure 7.2 showing a flood mark on a fully grown *Faidherbia albida* tree. With the dams, such type of floods may never happen.

Figure 7.2: Mark of a natural high flood level on *Faidherbia albida* tree. The top mark is about double the height of the right hand figure holding a vertical stick to indicate it (Attwell, 1970).

In order to keep the animals away from the floodplain for a sufficiently long period, floodplain flooding by a moderate flood of 5,800 m³/s appears enough (Figure 7.3). This size of this flood would be achieved through the opening of 3 Kariba spillway gates. With this moderate flood discharge, the lower terraces of the floodplain, as well as the old abandoned floodplain river channels, would be flooded, cutting off accessibility to most parts of the floodplain. The flood should be sustained for a period of 5 to 6 weeks in the months of February and March. It is in this period of the year that the local catchment streams are expected to also have peak flows, so that the tributaries would also contribute to the flooding (Figure 7.4 showing Mana River at peak flow, overflowing the Mana Bridge, February 2011). The water balance carried out for March 2011 (Sub-section 4.1.2.13) shows a contribution of about 400 m³/s from the tributaries, the Kafue River being the largest one. Table 7.1 shows the floodplain width expected to be flooded at the different flood sizes.

Phasing of the spill way gates closing for 3 to 4 weeks, in Sub-section 7.2.2 will ensure that the flood and ground water table recession is slowed down to keep the floodplain wet enough until the months of May and June, considering the slow drainage of the flooded areas, as observed from past flood events. This statement is based on field observations and experience of park operators, who were asked during the field campaigns of 2010 and 2011.

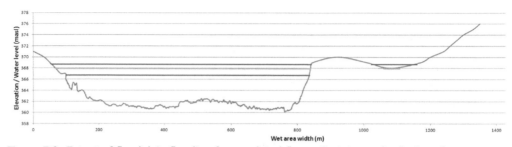

Figure 7.3: Extent of floodplain flooding for regulated flows: Qminimum (red), Qmoderate (green) and Qmaximum (purple) floods, cross-section at Chakanaka Farm (ZM3) (Figure 4.40 modified).

Table 7.1: Discharge with corresponding water levels and floodplain width (derived from Table 4.12).

Discharge	Chakanaka Farm		Chiawa Camp		Chikwenya island	
	Water level (masl)	Floodplain width (m)	Water level (masl)	Floodplain width (m)	Water level (masl)	Floodplain width (m)
Qminimum	366.7	744.0	351.0	1673.0	348.6	1859.3
Qmoderate	367.9	854.0	351.9	1764.0	349.1	2679.2
Qmaximum	368.7	884.5	352.5	1995.0	349.6	3233.0

Where
Qminimum = 4,500 m^3/s (2 spillway gates open)
Qmoderate = 5,800 m^3/s (3 spillway gates open)
Qmaximum = 7,000 m^3/s (4 spillway gates open)

Figure 7.4: Mana River in flood, overflowing over the Mana Bridge, February 2011.

7.2.1.2 *Other tree saving initiatives*

The proposed moderate flood will not be able to affect the higher terraces of the floodplains, therefore to save the *Faidherbia albida* tree from being exterminated in these areas, a deliberate "steel cage tree" (Figure 7.5 showing an example of the efforts to replant tree) planting during the rainy season is proposed. The steel cages will protect the young trees from being eaten by the animals until they are well established. This exercise should be done on an annual basis. This exercise should be undertaken by the Game Management Authorities in both Zambia and Zimbabwe. In addition, all the Lodge owners should be encouraged to plant the caged trees around their lodges. This program of caged tree planting should be extended to schools in the Game Management Areas of Zambia.

Figure 7.5: Past National Parks initiatives for replanting of the *Faidherbia albida* tree.

7.2.2 Saving the channels from the excessive bank erosion phenomenon

The fact that the aquifer hydraulic heads are higher than the river hydraulic heads means that a rapid water level drawdown would increase significantly the hydraulic gradient, unbalancing seepage forces and affecting the stability of the sandy banks of the Zambezi River. To mitigate the channel widening caused by the sudden closure of flood gates leading to the sudden reduction of water levels for about 5 meters within 24 hours (Sub-chapters 4.1.1.2 and 4.3.3), the phasing of the closure of spillway gates over a period of 3 to 4 weeks is proposed.

7.2.3 Monitoring Program

To implement the recommended measures without a monitoring program would not produce the intended benefits, since updating and monitoring may lead to fine tuning of the recommended measures. The flood releases would cost the hydropower operators, therefore, it is important that the effectiveness is monitored. The best recommended measures to monitor the flood extent would be to use the satellite images. This would be purchased and processed as part of the monitoring program. In addition, it is recommended ground truthing be carried out by a specialised team to ensure success of the measures. A thorough field campaign should be carried out after every moderate flood spill, which means every 5 years.

The success, expressed in quantitative terms of new *F. albida* establishments, could be best assessed during the months of July or August, considering that the moderate flood spill should occur in February-March and that the channels and parts of the floodplains will be flooded for a couple of months after complete closure of the spillways. Finally, hydropower operators should maintain the monitoring network that was setup during this research work (Sub-chapter 4.1.1) and collect regular hydraulic data.

7.2.4 Institutional arrangement and funding

As outlined in Sub-chapters 2.5.1 and 2.5.2, there are two important regional initiatives that would be best placed for anchoring the implementation of the recommended measures. At the Zambezi Watercourse Commission, would be mainly to ensure the measures are included in regional policies. The actual implementation would be by the Zambezi Water resources Managers and dam operators Committee. The funding for the implementation of the measures would be channeled to the Zambezi River Authority who have the mandate to manage the Zambezi River in the river reach shared between Zambia and Zimbabwe. The details of funding and monitoring program to be implemented by the Zambezi River Authority, should be designed and worked out by the Zambezi Water resources Managers and Dam Operators Committee (JOTC). It is therefore recommended that reports on the implementation and monitoring of the recommended measures for the Middle Zambezi becomes one of the agenda items during the meetings of the JOTC.

7.3   Conclusions

The Environmental flow regime and other initiatives recommended are mainly targeting at ensuring the sustainability of this important habitat for the purposes of the riparian tree survival and mitigating the artificially induced excessive channel bank erosion. Since the area is a hydropower dominated environment, there is need to change the focus of by ensuring that the downstream environment is also managed just as the reservoirs are taken care of.

The success of the implementation of the measures should be carefully monitored to serve as an encouragement for the effort being born by the dam operators. This may also serve as a learning curve for other areas that do need attention in the region.

It is imperative that research is supported and continues in the area, to ensure the fine-tuning process of the recommended measures.

CHAPTER 8: SYNTHESIS, CONCLUSIONS AND RECOMMENDATIONS

8.1   Synthesis

Since hydropower development was the major focus in the development stages of the Kariba and Kafue hydropower schemes on the Zambezi River, there was lack of attention to the environment downstream of these infrastructure, giving an impression that there is nothing much in the Middle Zambezi Valley to warrant investment into research and data collection. However, the middle Zambezi floodplains scored quite highly in terms of vegetation, mammals, birds, herps (reptiles and amphibians), and fish species in the Timberlake, (1998) assessment of wetlands of the Zambezi.

From the rich biodiversity that Middle Zambezi floodplains support, more attention should be given to the area. This research work has shown that there are hydrodynamic, morphodynamic and vegetation changes occurring as a result of the dam operations. The observed impacts need to be mitigated to ensure sustainability of this important area. The proposed mitigation and conservation measures would require deliberate trade-offs and funding that are justifiable, considering the rich biodiversity that makes this habitat special.

8.2   Discussion and conclusions

This study has provided answers to the following general questions:

> What is the current state of the Middle Zambezi River reach and its floodplains in view of the present flow regulation from the Kariba and Kafue hydropower schemes?

The research work included the analysis of the current dam operations and the response of the Middle Zambezi River reach and its floodplains. The analysis of the flow hydrographs show two distinct response periods: the post-Kariba1, being the time when there was one power station operating and the post-Kariba2, being the current state, showing the cumulative response of three power stations (two at Kariba and one at Kafue). The current state shows changes with respect to the pre-impoundment period that may not be reversible and some that can be mitigated by a change in dam operating rules (Chapter 4).

The impact of the hydropower regulation on the downstream environment has led to some permanent changes which may be difficult to reverse, showing that the study area is now hydropower-dominated and most of the flows in the river reach are regulated by the electricity needs. The following flow characteristics would be difficult to reverse:
- Due to dam operations the water levels significantly vary on the 24-hour circle, leading to a constant diurnal water level variation within a range of 15

cm corresponding to the time of the day and the power generation needs (Sub-chapter 4.1.1.1).

- The dry season flows have increased from 100 m^3/s to over 1,000 m^3/s. This constant flow of over 1,000 m^3/s is due to the turbine outflows from Kariba and Kafue Power Stations (Sub-chapter 4.2.3.3).

Changes that can be mitigated and may need an effort in policy shift in the dam operating rules are:

- The complete absence of flood flows for extended periods (Sub-chapter 4.2.3.2).
- The extremely short flood spillage periods characterised by extreme (up to 5 metres) water level fluctuation during a 24 hour period due to the opening and closing of the dam spillway gates (Sub-chapter 4.1.1.2). This particular change presents an extreme departure from the natural system and may resulting in negative consequences in terms of ecology and morphology.

What was the state of the river reach and its floodplains before Kariba and Kafue hydropower schemes were constructed?

The analysis was done with the help of a hydrodynamic model, where unregulated flows were used. The results show that the river reach was able to benefit from a predictable natural rhythm of annual increase and recession of flows depending to annual rainfall, with annual minimum flows ranging between 100 to 300 m^3/s. It also shows that floods occurred at irregular intervals of years with high annual peak flows as follows:

- (>8,000 m^3/s) - extreme high floods;
- (6,000 to 7,000 m^3/s) high floods;
- (4,000 to 6,000 m^3/s) moderate floods; and
- (3,000 to 4,000 m^3/s) low floods.

Before dam construction there were also dry periods, in which the discharge was less than 2,500 m^3/s, when there was no flooding (Sub-chapter 4.2.3.1). From the morphodynamic analysis (Chapter5), the results show the river reach with a lower degree of braiding with respect to present, with larger sand bars and islands in most sections of the floodplain (Sub-chapter 5.2.1 and 5.2.2).

> What is the difference between the past and the current state of the middle Zambezi River reach and floodplains?

In terms of hydrodynamics (Sub-chapter 4.2.3.3), the following can be deduced:

- The seasonal low flows have become higher, being about 1,000 m^3/s up to 2001, and above 1,000 m^3/s and rising for the period from 2001. The increase can be attributed to the full operation of all the three power stations and a general increase in electricity demand in the region. In the natural system, the low flows were less that 100 m^3/s in the dry season.
- From a more predictable rhythm of flow peaking and recession, peak flows are now of short duration, irregular and scarce and totally dependent on flood spillages. Only 4 short spillage flows have occurred in a period of 22 years (1993 to 2013).

The stable and more predictable morphodynamics of the pre-dam state have been overtaken by significant bank failure leading to channel widening and eroding of sand bars and islands due to the dam operations (Sections 5.2.1 and 5.2.2).

> What is the role of the tributary streams in discharge and sediment supply to this river reach?

Although the Middle Zambezi River reach lies within a low rainfall area, with mean annual rainfall of 650 mm, characterised by a short rain season from December to mid March and prone to droughts, it is expected that in the months when the local catchment area experiences a good rainfall distribution, the local tributaries would be able to make some significant contribution to discharge generation within the catchment. This has been quantified in 400 m^3/s (Sub-chapter 4.1.2.13). The morphodynamic research shows that the Middle Zambezi tributaries still carry significant amount of sediments both in terms of suspended and bed sediments (Sub-chapter 5.1.2). The results show that although the hydropower infrastructure and associated operations have had an impact on the morphodynamic processes of the Middle Zambezi River reach, the local catchment area and its tributaries have continued to contribute significantly to the supply and distribution of sediments, which is key to the morphodynamic processes of this river reach.

> What are the interactions between floodplain vegetation, surface and subsurface flows?

The research work used the *Faidherbia albida* tree, which constitutes a key supply of the much-needed nutrients during the dry season when there is a shortage of food, to study the effects of flow regulation on the floodplain forest. During the field surveys it was observed that in the higher floodplain terraces the tree distribution is characterised by a stand of old trees (about 100 years old) and lack of tree regeneration coupled with a dying-off of older trees, painting a gloomy picture of a once thriving habitat experiencing changes that may drastically change the quality of the habitat. Younger trees were only observed to thrive on sand bars and islands that are less accessible to the animals. The results show that the survival of this floodplain tree is highly dependent on the restoration of the occasional floodplain flooding as it used be before the dams were constructed. In the past, the floods were able to keep away the animals from some sections of the floodplain to aid the critical regeneration of this important tree. In the absence of flooding, the residence time of the animals on the floodplain has much increased, making the tree regeneration close to nothing, because all the seedlings are eaten as soon as the seeds sprout (Sections 4.2.3.2, 6.1.3, 6.2.2 and 6.4).

The research on the subsurface flows showed that the regional aquifer on the floodplains drains towards the Zambezi River. This is a very important factor from the river morphology point of view as it has severe consequences for bank erosion. The fact that the aquifer hydraulic heads are higher than the river hydraulic heads means that a rapid water level drawdown would increase significantly the hydraulic gradient, unbalancing seepage forces and affecting the stability of the sandy banks of the Zambezi River. This reinforces the need to make a shift in the way the flood gates are opened and closed to save the downstream river channels from excessive bank erosion (Sub-chapter 4.3). The research has also shown that the flow regulation has not substantially affected the aquifer hydraulic heads. Therefore the observed floodplain forest degeneration cannot be attributed to changes in aquifer heads.

> What will be the state of river reach and floodplains in the future if the current water regulation remains the same?

The river reach being so dependent on the outflows from the dams means that the general decline of the quality of the floodplains, in particular as relates to the survival of the riparian forest tree *Faidherbia albida*, will continue. This will lead to the complete disappearance of the *Faidherbia albida* tree from this habitat, rendering it unable to sustain the rich population of wildlife that depend on the tree for sustenance, especially during the critical periods of food shortage in the dry season. This will have an effect on the socioeconomic activities that are associated to wildlife management. The tree, with its many beneficial characteristics and food value, appears as the resource that keeps the animals on this floodplain, making the area popular for tourism activities. The sandbars and islands have been observed as areas where the tree is still flourishing and if bank failure and channel widening phenomena continue, these areas may gradually reduce in size. This would lead to the loss of the its scenic beauty as more and more morphorhological features are lost to the high rate of channel widening phenomenon. The high rate of channel widening may therefore lead to loss of both vegetation and tourism infrastructure.

> What is the environmental flow regime that is required to minimise the impact of the upstream hydropower schemes?

The survival of the *Faidherbia albida* tree is symbiotic to the quality of the habitat. The tree is the supplier of the much-needed food for wildlife during the critical dry months of the year, when the balk, the leaves and the seed are all eatable. This is critical food for the animals providing the much-needed nutrients to keep them healthy. Therefore, just as it is hold under protection in semi-arid regions where it grows, this tree needs to be conserved and preserved in the Middle Zambezi (Chapters 6 and 7). To save the *Faidherbia albida* tree from being completely exterminated from this floodplain, by keeping the animals away from some sections of the floodplain for a period that is long enough, a two pronged environmental flow regime is proposed as follows:

- Once every five years, the dam operators should facilitate a deliberate release of a moderate flood of 5,800 m^3/s which should have duration of 5 to 6 weeks in the months of February to March.
- The spill way gates closing should be phased for 3 to 4 weeks, to ensure that the flood and ground water table recession is slowed down to keep the floodplain wet enough until the months of May and June. This would allow for germination and regeneration of the seedlings to be established (Chapter 7).

The proposed moderate flood will not be able to affect the higher terraces of the floodplains, therefore to save the *Faidherbia albida* tree from being exterminated in these areas, a deliberate "steel cage tree" planting during the rainy season is proposed. The steel cages will protect the young trees from being eaten by the animals until they are well established. This exercise should preferably be done jointly on an annual basis by the Game Management Authorities in both Zambia and Zimbabwe. In addition, all the lodge owners should be encouraged to plant the caged trees around their lodges. This program of caged tree planting should be extended to schools in the Game Management Areas of Zambia (Chapter 7). To mitigate the channel widening caused by the excessive bank erosion as a result of sudden closure of flood gates leading to the sudden reduction of water levels for about 5 meters within 24 hours, the phasing of the closure of spillway gates over a period of three to four weeks is proposed (Chapter 7).

8.3 Recommendations

The need for the hydropower operators to start caring for the downstream environment as part of the hydropower schemes needs a policy shift. The Middle Zambezi being a shared resource between Zambia and Zimbabwe, it is recommended that the measures recommended in Sub-chapter 8.2 should be presented to the Zambezi River course Commission (Sub-chapter 2.5.1) for adopting. These should then be scaled down to the Zambezi Water Resources Managers and Dam Operators Committee for implementation (Sub-chapter 2.4.2).
The measures recommended would need funding to succeed, therefore, just as the dam operations and maintenance is part of the budgets of the hydropower operators, it is recommended that the management and conservation of the

downstream environment be included in the mainstream budgets for the hydropower operators. At least 10% of the annual dam operations and maintenance should be dedicated to the conservation and monitoring efforts of the downstream environment. This will be much more sustainable, than leaving the recommendations hanging with no one in particular to hold accountable.

It is further recommended that monitoring programs for the downstream environment be included to ensure that such a significant habitat is conserved and sustained. In particular, regular monitoring will allow fine-tuning of the proposed environmental flow regime, based on its efficacy in guaranteeing the establishment of new *F. albida* trees. To facilitate the success of the monitoring programme of the downstream environment, it is recommended that the hydropower operators establish and maintain the monitoring network during this research work (Sub-chapter 4.1.1). Consideration for installation of cost-effective devices such as divers can be made whose data can be downloaded at intervals to begin building a database of observed data for future studies. A thorough field campaign should be carried out after every moderate flood spill, which means every 5 years. The success, expressed in quantitative terms of new *F. albida* establishments, could best be assessed during the months of July or August.

References

Abd El-Aziz T. M and N. M. Abd El-Salam. 2007. *Characteristic Equations for Hydropower Stations of Main Barrages in Egypt.* In proceedings of the Eleventh International Water Technology Conference, IWTC11 2007 Sharm El-Sheikh, Egypt. (pp 461-470).

Acreman MC, Ferguson AJD. 2010. *Environmental Flows and the European Water Framework Directive.* In Journal of Freshwater Biology Vol 55: pp32–48. Wiley Online Library.

Akehurst A., Newley P and M. Hickey, 2008. *Soil and Water best Management for NSW Banana Growers.* NSW Department of Primary Industries. New South Wales.

Attwell R. I., 1970. *Some effects of Lake Kariba on the ecology of a floodplain of the Mid-Zambezi Valley of Rhodesia.* In Journal of *Biological Conservation*, Vol. 2 No. 3, April 1970: pp189-196. Elsevier Publishing Company Ltd, England. http://www.sciencedirect.com/science/article/pii/0006320770901060

Axelrod D. I and P. H. Raven 1978. *Late Cretaceous and Tertiary vegetation history of Africa.* In Merger M. J. A and A. C. Van Bruggen (eds). Journal of Biogeography and Ecology of Southern Africa, W. Junk, The Hague.

Barnes, R.D. and Fagg, C.W., 2003. *Faidherbia albida monograph and annotated bibliography.* Tropical Forestry Papers No 41, Oxford Forestry Institute, Oxford, UK.

Basson, G., 2005. *Hydropower Dams and Fluvial Morphological Impacts An African Perspective.* Department of Civil Engineering, University of Stellenbosch, South Africa. In Paper from United Nations Symposium on Hydropower and Sustainable Development of 2004.(pp. 27-29).

Beadle, L. C., 1982. *The Inland Waters of Tropical Africa.* In the 2nd Edition, Longman, London, UK.

Beilfuss, R., 2012. A Risky Climate for Southern African Hydro. ASSESSING HYDROLOGICAL RISKS AND CONSEQUENCES FOR ZAMBEZI RIVER BASIN DAMS. International Rivers, 2150 Allston Way, Suite 300, Berkeley, CA 94704, USA.

Bunn, S.E. & Arthington, A.H. 2002. Basic principles and ecological consequences of altered flow regimes for aquatic biodiversity. *Environmental Management* 30 (4): 492–507.

Boffa J. M., 1999. Agroforestry parklands in Sub-Saharan Africa. Forestry Department. FAO Conservation Guide 34. FAO Corporate Document Depository. Rome, Italy. M-36 ISBN 92-5-104376-0. http://www.fao.org/docrep/005/x3940e/x3940e00.htm

Bond G., 1975. The Geology and Formation of the Victoria Falls. In Phillipson D. W (ed), Mosi-ao-Tunya: A Handbook to the Victoria Falls Region. Salisbury, Rhodesia.

Brown C and KING J., 2011. Modifying dam operating rules to deliver environmental flows: experiences from southern Africa. Intl. J. River Basin Management iFirst, 2011, 1–16. International Association for Hydro-Environment Engineering and Research. http://dx.doi.org/10.1080/15715124.2011.639304 / http://www.tandfonline.com

Brown C. and P. Watson, 2007. Decision support system for environmental flows: Lessons from Southern Africa. International Journal of River Basin Management Volume 5, No. 3.

Bunn, S.E. & Arthington, A.H. 2002. Basic principles and ecological consequences of altered flow regimes for aquatic biodiversity. *Environmental Management* 30 (4): 492–507.

Brown, C. & J. King (2011): Modifying dam operating rules to deliver environmental flows: experiences from southern Africa, International Journal of River Basin Management, DOI:10.1080/15715124.2011.639304. http://dx.doi.org/10.1080/15715124.2011.639304

Centre technique forestier tropical (CTFT). 1988. Faidherbia albida *(Del.) A. Chev. (Synonyme:* Acacia albida *Del.). Monographie.* Nogent-sur-Marne, France, CTFT. 72 pp. http://core.ac.uk/download/files/449/12103007.pdf

Crosato, A., 2014. Environmental Flows Assessment Geomorphology Issue. Environmental Water Allocation. Lecture Notes (LN0454). UNESCO-IHE, Delft, The Netherlands.

Cumming D. H. M., 1983. Editorial - Zambezi Valley Ecosystem. In The Zimbabwe Science News. Volume 17 (7/8): pp 120.

Davies B. R., 1986. The Zambezi River. In Davies B. R and K. F. Walker (eds), The Ecology of River Systems. Dr. W. Junk Publishers, Lancaster, UK.

Delft Hydraulics, 2006. DELFT3D-FLOW User manual, Deltares, Delft, the Netherlands. http://www.deltaressystems.com/hydro/product/621497/delft3d-suite).

Deltares, 2011. Tutorial SOBEK-Rural FLOW Module Workshop, Ho Chi Minh City June 1-3, 2011. Deltares, Delft, The Netherlands.

Deltares . SOBEK online help. Deltares, Delft, Utrecht, The Netherlands
http://www.deltares.nl/nl/software/108282/sobek-suite
http://www.deltares.nl/en/software/108282/sobek-suite/1168665.

Department of Water and Sanitation (DWS). 2014. Resource Directed Measures:
Reserve determination study of selected surface water and groundwater
resources in the Usuthu/Mhlathuze Water Management Area. Pongola
Floodplain – EWR Report. Report produced by Tlou Consulting (Pty) Ltd.
Report no: RDM/WMA6/CON/COMP/1113.

DHV Consultants, 2004. Decision Making System for Improved Water Resource
Management for the Kafue Flats – Integrated Water Resources Management
Project for the Kafue Flats Phase 2. WWF, Lusaka, Zambia.
http://wwf.panda.org/who_we_are/wwf_offices/wwf_zambia_nature_conserv
ation/

Dunham K. M., 1989. Vegetation-environment relations of a Middle Zambezi
floodplain. In Vegetation Vol. 82 (pp13-24). Kluwer Academic Publishers,
Belgium.

Dunham K. M., 1990a. Biomass dynamics of herbaceous vegetation in Zambezi
Riverine woodlands. In African Journal of Ecology, Volume 28, pp. 200-212.

Du Toit R.F., 1982. A preliminary Assessment of the Environmental Implications of
the Proposed Mutapa and Batoka Hydro-electric Schemes (Zambezi River,
Zimbabwe). Natural Resources Board, Harare, Zimbabwe.

Du Toit R. F., 1983. Hydrological Changes In the Middle Zambezi System. Zimbabwe
Science News 17(7/8):.121-125.

Du Toit R. F., 1984. Some environmental aspects of proposed hydro-electric
schemes on the Zambezi River, Zimbabwe. In Journal of Biological
Conservation, Volume 28, pp. 73-87.

Ekandjo M., 2011. Hydrological analysis of the Middle Zambezi and impacts of the
hydropower dams on the flow regime in Mana Pools National Park. (MSc.
thesis). University of Zimbabwe, Harare, Zimbabwe.

Engelund F. & Hansen E., 1967. *A Monograph on Sediment Transport in Alluvial
Streams,* Teknisk Forlag. Copenhagen-V, Denmark

FAO 1977, *Report on the 4th Session of the FAO Panel of Experts on Forest Gene
Resources.* FO:FGR/4/Rep. Rome. FAO.
http://www.fao.org/docrep/006/ad449e/AD449E00.htm

Gope E. T., 2012. *Effects of flow alteration on Faidherbia albida stands of the Middle
Zambezi Floodplains.* (MSc. Thesis). UNESCO-IHE, Delft, The Netherlands.

Gope, E. T., Sass-Klaassen, U. G. W., Irvine, K., Beevers, L., & Hes, E. M. A. 2015. *Effects of flow alteration on Apple-ring Acacia (Faidherbia albida) stands, Middle Zambezi floodplains, Zimbabwe.* Journal of Ecohydrology, Volume 8(5), pp 922–934. 10.1002. DOI: 10.1002/eco.1541.

Gordon, R. (1989). Acoustic Measurement of River Discharge. In Journal of Hydraulic Engineering: pp 925-936., doi: 10.1061/(ASCE)0733-9429(1989)115:7(925).

Gordon N. D., McMahon T.A and Finlaysson B. L., 1992. In Stream Hydrology. An introduction for ecologists. John Wiley & Sons Ltd, Chichester, England.

Grissino-Mayer H. D., 2003. A manual and tutorial for the proper use of an increment borer. In Tree-Ring Research volume 59(2): pp 63-79.

Gupta, R. S., 1989. *Hydrologic and Hydraulic Systems,* Prentice Hall, Eaglewood Cliffs, New Jersey, pp. 53 - 55.

Gurnell, A.M., Bertoldi,W., and Corenblit, D. (2012). Changing river channels: The roles of hydrological processes, plants and pioneer fluvial landforms in humid temperate, mixed load, gravel bed rivers. *Earth-Science Reviews* 111(1-2): 129–141.

Guy, P.R. (1981). River bank erosion in the Middle Zambezi Valley, downstream of Lake Kariba. Biological conservation 19: 119-212.

GWP(Global Water Partnership)/INBO (International Network of Basin Organizations), 2009. *Handbook for IWRM in Basins.* Elanders, Sweden. http://www.unwater.org/downloads/gwp_inbo%20handbook%20for%20iwrm%20in%20basins_eng.pdf

Heuzé V., Tran G., 2013. *Apple-ring acacia (Faidherbia albida).* Feedipedia.org. A programme by INRA, CIRAD, AFZ and FAO. http://www.feedipedia.org/node/357 (Last updated on June 21, 2013, 8:26)

Hoehn E., 1998. *Solute exchange between river water and groundwater in headwater environments. Hydrology, Water Resources and Ecology in Headwaters.* in Proceedings of the Headwater 1998 Conference held at Meran/Marano, Italy, April 1998). IAHS Publ. No. 248, 1998.

Hooke JM. 1979. An analysis of the processes of riverbank erosion. Journal of Hydrology 42: 39–62.

Hughes D.A. 2001 *Providing hydrological information and data analysis tools for the determination of ecological instream flow requirements for South African rivers.* In Journal of Hydrology, Volume 241, p. 140–151.

Hughes F. M. R. 1988. The ecology of the African Floodplain Forests in semi-arid and arid zones: a review. In Journal of Biogeography (1988), volume 15, pp 127-140

Hughes F. M. R. 1990. The influence of flooding regimes on forest distribution and composition in the Tana River Floodplain, Kenya. In Journal of Applied Ecology (1990). Volume 27 (pp 475-491)

Hughes F. M. R. 1994. Environmental Change, Distribution and Regeneration in Semi-arid Floodplain Forests. In Environmental Change in Drylands: Biogeographical and Geomorphological Perspectives. Millington A. C and K. Pye (eds). John Wiley & Sons Ltd.

Hughes F. M. R. 1997. Floodplain biogeomorphology. In *Progress in physical geography*, volume *21*(4), pp.501-529. Hodder Headline Group, London, UK. ISSN 0303-13333.

Hughes F.M.R., del Tánago M. G. and Mountford J. O. 2012. Restoring Floodplain Forests in Europe (pp 393-422). In J. Stanturf et al. (eds.), *A Goal-Oriented Approach to Forest Landscape Restoration*, World Forests 16, DOI 10.1007/978-94-007-5338-9_15, © Springer Science+Business Media Dordrecht 2012

IUCN, 1996. Zambezi basin wetlands conservation and resource utilisation project. Inception mission report, June 1996. IUCN ROSA, Harare, Zimbabwe.

Jackson P. B. N., 1986. *Fish of the Zambezi system*. In Davis B. R and K. F. Walker (eds), The Ecology of River Systems. Dr. W. Junk Publishers, Lancaster, UK.

Jain Sharad K. 2012. *Assessment of environmental flow requirements*. In HYDROLOGICAL PROCESSES (2012). Wiley Online Library (wileyonlinelibrary.com). Doi: 10.1002/hyp.9455

Jansen P. P., Bendegom, L. V. D., Berg, J. v. d., Breusers, H. N. C., and Dekker, J., 1979, *Principles of river engineering: the non-tidal alluvial river*. Pitman, London. (Repository TU Delft)

Jarman P. J., 1972. Seasonal Distribution of large Mammal population in the unflooded Middle Zambezi Valley. In Journal of Applied Ecology, Volume 9 No. 1 of 1972. (pp 283-299). British Ecological Society. http://www.jstor.org/stable/i317629

Jensen M. E., 2010. *Estimating Evaporation from water surfaces* - Extracts from Chapter 6 of the ASCE Manual 70, second edition (pp 1-27). Fort Collins. Colorado, USA.

Jones F.E., 1992. *Evaporation from water: with emphasis on application and measurement*. Lewis Publishing Inc., Chelsea, Michigan, USA.

Jubb R. A., 1967. Freshwater Fisheries of the Southern Africa. Published by A. A. Balkema, Cape Town, South Africa.

Khan Omar, 2013. Effects of dam operations on the Mana Floodplain forest in the Middle Zambezi - The role of flow regime, river morphology and groundwater. (MSc. thesis). UNESCO-IHE, Delft, The Netherlands.

Khan O., Mwelwa-Mutekenya E., Crosato A., Zhou Y., 2014. Effects of dam operation on downstream river morphology: the case   of the middle Zambezi River. Proceedings of the Institute of Civil Engineers (ICE). Volume 167, Issue 10, November 2014, p. 585-600. Doi: 10.1680/wama.13.00122.

Kohler M. A., Nordenson J. J., and Fox W. E., 1955. *Evaporation from pans and lakes*. US Department Comm., Weather Bur. Res., Paper 38, 21 p.

Kondolf, G.M. (1997). Hungry water: Effects of dams and gravel mining on river channels. Environmental Management 21(4): 533–551.

Kondolf, G.M., Gao, Y. et al. (2014), Sustainable sediment management in reservoirs and regulated rivers: Experiences from five continents. *Earth's Future* 2: 256–280, doi:10.1002/2013EF000184.

Kondolf, G.M. and Wilcock, P.R. (1996). The flushing flow problem: Defining and evaluating objectives. *Water Resources Research* 32(8): 2589-2599.

King L. C., 1978. The Geomorphology of Central and Southern Africa. In Werger M. J. A and A. C. Van Bruggen (eds), Biogeography and Ecology of Southern Africa. W. Junk, The Hague, The Netherlands.

Krchnak K, Richter B, Thomas G. 2009. Integrating EFs into Hydropower Dam Planning, Design, and Operations. In Water Working Note No. 22, World Bank, Washington DC., USA.

Kumar, A., T. Schei, A. Ahenkorah, R. Caceres Rodriguez, J.-M. Devernay, M. Freitas, D. Hall, A. Killingtveit, Z. Liu, 2011: Hydropower. In Edenhofer, R. (ed) IPCC Special Report on Renewable Energy Sources and Climate Change Mitigation.

Kunz, M.J. (2011). Effect of large dams in the Zambezi River Basin: changes in sediment, carbon and nutrient. PhD Thesis, No. 19441, ETH Zurich, Switzerland.

LeHouerou, H., 1980b. The Role of Browse in the Sahhelian and Sudanian Zones Zones; Chemical Composition and Nutritive Value of Browse in West Africa. Both in ILCA, Browse in Africa: The Current State of Knowledge, pages 83-102 and 261-290. Addis Ababa, Ethiopia.

Lu X. X and R. Y. Siew, 2006. Water discharge and sediment flux changes over the past decades in the Lower Mekong River: possible impacts of the Chinese dams. Hydrol. Earth Syst. Sci., Volume 10, p. 181–195.

Ma, Y., Huang, H.Q., Nanson, G.C. , Li , Y. and Yao W. (2011). Channel adjustments in response to the operation of large dams: The upper reach of the lower Yellow River. Geomorphology 147-148: 35-48. doi:10.1016/j.geomorph.2011.07.032

Malvern Viscount 1960. Kariba - The story of the world's biggest man-made lake. Central News Agency Limited, Bloemfontein, South Africa.

Matos J.P., T. Cohen, J.-L. Boillat, A.J. Schleiss, M.M. Portela, 2010. Analysis of flow regime changes due to operation of large reservoirs on the Zambezi River. In Christodoulou and Stamou (eds).Proceedings of the 6th International Symposium on Environmental Hydraulics held in Athens, Greece. Taylor & Francis Group, London, ISBN 978-0-415-58475-3.

McCartney, M. P. 2007. *Decision support systems for large dam planning and operation in Africa.*: International Water Management Institute. 47 p. (in IWMI Working Paper 119). Colombo, Sri Lanka. ISBN 978-92-9090-662-9 Copyright © 2007, by IWMI (*iwmi@cgiar.org) /* http://www.iwmi.org

Mc Clain M.E., 2013. Balancing Water Resources Development and Environmental Sustainability in Africa: A Review of Recent Research Findings and Applications. Springerlink.com www.kva.se/en 123 AMBIO 2013, Volume 42, p.549–565. Doi: 10.1007/s13280-012-0359-1

Mertens Jasmin (ed), 2013. African Dams Project. An Integrated Water Resources Management Study. Final Stakeholder Report. CCE / ADAPT, 2013. ETH, Zurich, Switzerland. The ADAPT website: http://www.cces.ethz.ch/projects/nature/adapt

MEWD, WWF and ZESCO, 2003. Tripartite agreement for the Intergrated Water Resources Management of the Kafue Flats. WWF, Lusaka, Zambia.

Mubambe C. G., 2012, Effects of Large Dams on Riverine Geomorphology and Riparian Vegetation Case Study of Mana Pools Floodplains - Middle Zambezi River (MSc thesis). University of Zimbabwe, Harare, Zimbabwe.

Mukono, S.N. (1999). Predicting induced long-term morphological changes on the Zambezi River due to hydropower projects. The Kariba Dam case. In Proceedings Waterpower'99: Hydro's Future: Technology Markets and Policy, Edited by Brookshier P.A., Chapter 8, ASCE, http://ascelibrary.org/doi/abs/10.1061/40440%281999%298

National Institute of Hydrology, 1999. *Estimating Hydrologcal Parameters for Water Balance Studies in Tambaraparni River Basin,* Tamil Nadu, Jalvigyan Bhawan Roorkee, India.

Ncube S., 2011. The interaction of flow regime and the terrestrial ecology of the Mana Floodplains in the middle Zambezi River Basin.(MSc. Thesis). UNESCO-IHE, Delft, The Netherlands.

O'Keeffe Jay, Tom Le Quesne, 2009. Keeping Rivers Alive. A primer on environmental flows. WWF Water Security Series 2. http://assets.wwf.org.uk/downloads/keeping_rivers_alive.pdf

Orwa, C., Mutua, A., Kindt, R., Jamnadass, R. and Simons, A., 2009. Agroforestree Database: a tree reference and selection guide version 4.0. World Agroforestry Centre. Nairobi, 00100, Kenya :worldagroforestry@cgiar.org www.worldagroforestry.org

Palmberg C (1981). A vital fuel wood gene pool is in danger. Unasylra, 133: 22-30. http://www.fao.org/docrep/p3350e/p3350e04.htm

Panday Sorab, Langevin, C.D., Niswonger, R.G., Ibaraki, Motomu, and Hughes, J.D., 013, MODFLOW-USG version 1: An unstructured grid version of MODFLOW for simulating groundwater flow and tightly coupled processes using a control volume finite-difference formulation: U.S. Geological Survey Techniques and Methods, book 6, chap. A45, 66 p. U.S. Geological Survey, Reston, VA. USA.

Parker G., Shimizu Y., Wilkerson G. V., Eke E. C., Abad J. D., Lauer J. W., Paola C., Dietrich W. E. and Voller V. R., 2011. A new framework for modeling the migration of meandering rivers. Earth Surf. Process. Landforms, Volume 36, p. 70–86. doi: 10.1002/esp.2113.

Petts, G.E., Gurnell, A.M., 2005. Dams and geomorphology: research progress and future directions. Geomorphology 71, 27–47. DOI : 0.1016/j.geomorph.2004.02.015

Pinhey E. C. G., 1978. Odonata. In Werger M. J. A and A. C. Van Bruggen (eds), Biogeography and Ecology of Southern Africa. W. Junk, The Hague, The Netherlands.

Prinsen G. F. and Becker B. P. J., 2011. Application of SOBEK hydraulic surface water models in the Nitherlands Hydrological Modelling Instrument. Irrigation and Drainage 60 (Suppl. 1), p. 35-41.

Richter, B.D. & Thomas, G.A. 2007. Restoring environmental flows by modifying dam operations. *Ecology and Society* 12 (1).

Richter, B.D. & Thomas, G.A. 2007. Restoring environmental flows by modifying dam operations. *Ecology and Society* 12 (1).

Rinaldi, M., and Casagli, N. (1999). Stability of strembanks formed in partially saturated soils and effects of negative pore water pressure: the Sieve River (Italy). Geomorphology 26(4): 253-277.

Rinaldi, M., Casagli, N., Dapporto, S. and Gargini, A. (2004). Monitoring and modelling of pore water pressure changes and riverbank stability during flow events. Earth Surf. Process. Landforms 29: 237–254.

Rinaldi, M., Mengoni, B., Luppi, L., Darby, S. E. & Mosselman, E. 2008. Numerical simulation of hydrodynamics and bank erosion in a river bend. *Water Resources Research,* Volume **44,** W09428. doi:10.1029/2008WR007008. Wiley Online Library.

Ronco, P., Fasolato, G., Nones, M. et al. 2010. Morphological effects of damming on lower Zambezi River. In *Geomorphology.* Volume 115 (1-2) 2010, pp 43-55. Elsevier. http://www.sciencedirect.com/science/article/pii/S0169555X09004048

Rood S. B, Braatne J. H. and Goater L. A., 2010. Responses of obligate versus facultative riparian shrubs following river damming. River Research and Applications (2010). Volume 26: pp102–117. (www.interscience.wiley.com) DOI: 10.1002/rra.1246

Rood S. B., Samuelson G. M., Braatne J. H., Gourley C. R., Hughes F. M. R, and Mahoney J. M. Managing river flows to restore floodplain Forests. *Front Ecol Environ* 2005; 3(4): 193–201. The Ecological Society of America. www.frontiersinecology.org

Rood S. B., Gourley C. R., Ammon E. M., Heki L. G., Klotz J.R., Morrison M. L., Mosley D., Scoppettone G. G., Swanson S, And Wagner P. L. 2003. Flows for Floodplain Forests: A Successful Riparian Restoration. BioScience *2003, Volume. 53 No. 7.* pp 647- 656

SADC, 2004. The Agreement on the Establishment of the Zambezi Watercourse Commission. ZAMCOM Secretariat, Harare, Zimbabwe.

SADC-WD/ZRA., 2008. Integrated Water Resources Management Strategy and Implementation Plan for the Zambezi River Basin. SIDA / DANIDA, Nowergian Embassy, Lusaka, Zambia.

SAPP., 2014. Overview of Southern African Power Pool. A presentation at the South Asia Regional Workshop on Competitive Electricity Markets 18-20 March 2014. www.sapp.co.zw

Scott Wilson Piésold, 2003. Conjunctive Operation of Kafue and Kariba Hydropower Developments. Integrated Kafue River Basin Environmental Impact Assessment Study OPPPI/ZESCO, Lusaka, Zambia.

Simon, A. and Collison A.J.C. (2002). Quantifying the mechanical and hydrologic effects of riparian vegetation on streambank stability riparian vegetation and fluvial geomorphology. *Earth Surf. Process. Landforms* 27: 527–546 (2002), DOI: DOI: 10.1002/esp.325

Simon, A., Curini A., Darby S.E., Langendoen E.J. (2000).Bank and near-bank processes in an incised channel. *Geomorphology* 35 : 193-217.

Simon, A. and Darby S.E. (2002). Effectiveness of grade-control structures in reducing erosion along incised river channels: the case of Hotophia Creek, Mississippi. *Geomorphology* 42: 229– 254.

Simpson M., 2001. Discharge measurement using a broad-Band Acoustic Doppler Current Profiler. USGS Open-File Report 01-01, Sacramento, California, USA.

Skelton. P., 1994. Diversity and distribution of freshwater fishes in east and southern Africa. Annales de Musée Royal de l'Afrique Centrale, Zoologie, Volume 275, p. 95-131.

Soil Conservation Service, 1986. *National Engineering Handbook, Section 4. Hydrology.* Department of Agriculture, Washington.

Sokolov A. A and Chapman T. G., 1974. *Methods of Water Balance Computations. An international guide for research and practice.* The UNESCO Press, Paris, France.

SonTek YSI, 2010. RiverSurveyor S5/M9 System Manual Firmware Version 1.0. San Diego, USA. http://www.sontek.com.

Soulard F, 2003. *Water accounting at Statistics Canada: The inland fresh water assets account.* Statistics Canada, Ottawa, Ontario, Canada.

Thorne CR. 1982. Processes and mechanisms of river bank erosion. In Gravel-bed Rivers, Hey RD, Bathurst JC, Thorne CR (eds). Wiley: Chichester; 227–271.

Timberlake J., 1998. Biodiversity of the Zambezi Basin Wetlands. Review and preliminary assessment of available information - Phase1. IUCN - ROSA, Harare, Zimbabwe.

Timberlake J., 2000. Biodiversity of the Zambezi Basin. In Occasional Publications in Biodiversity No. 9. Biodiversity Foundation for Africa, Bulawayo, Zimbabwe.

United States Department of the Interior, 2000. Lower Colorado River Accounting System, Demonstration of technology, Bureau of Reclamation, Lower Colorado Regional Office. USA.

United States Department of the Interior, 2007. *Lower Colorado River Accounting System: Evapotranspiration and Evaporation calculations.* Calendar year 2005. Lower Colorado regional Office. Boulder City, Nevada.

UNESCO-IHE, 2009. Power2Flow, Hydropower-to-environment water transfers in the Zambezi basin: balancing ecosystem health with hydropower generation in hydropower-dominated basins. UNESCO-IHE Institute for Water Education, Delft, the Netherlands.

UNESCO World Heritage Centre 1992-2016. Mana Pools National Park, Sapi and Chewore Safari Areas. Urungwe District, Mashonaland North region. S15 49 10 E29 24 29. **Date of Inscription:** 1984. http://whc.unesco.org/en/list/302.

Vanon V. A., (ed), 1975. Sediment Engineering,. ASCE Manual and Reports on Engineering Practice., 54, New York, USA.

Wentling Mark. G., 1983. *Acacia albida:* Arboreal Keystone of Successful Agro-Pastoral Systems in Sudano-Sahelian Africa. Submitted to: Dr. R.E. McDowell International Professor Department of Animal Science, Cornell University Ithaca, New York 14853 - May 3, 1983.

White E.,1969. Man-made lakes in tropical Africa and their biological potentialities. Biological Conservation. Voluime 1, Issue 3, p. 219-24.

Wickens, G.E. 1969. A study of Acacia albida Del. (Mimosoideae). Kew Bulletin, Volume 23, Issue 2, p. 181–202.

Wickens, G., 1980. *Alternative Uses of Browse Species.* in ILCA, Browse in Africa: The Current State of Knowledge. pp 155 -184. Hourerou, H. N. (ed). Addis Ababa, Ethiopia.

Wilson, G.V., Periketi, R.K., Fox, G.A., Dabney, S.M., Shields, F.D. and Cullum, R.F. (2007).Soil properties controlling seepage erosion contributions to streambank failure. Earth Surf. Process. Landforms 32: 447–459. DOI: 10.1002/esp.1405

Wolman MG. 1959. Factors influencing the erosion of cohesive river banks. American Journal of Science 257: 204–216.

Worbes M., Staschel R., Roloff A., and W. J Junk. 2003. *Tree ring analysis reveals age structure, dynamics and wood production of a natural forest stand in Cameroon.* In Forest Ecology and Management. Volume 173, issues 1-3: pp 105-123. Elsevier. DOI: 10.1016/S0373-1127(01)00814-3

World Bank. 2010. The Zambezi River Basin *A Multi-Sector Investment Opportunities Analysis.* Volume 1, Summary Report. THE WORLD BANK Water Resources Management Africa Region. © 2010 The International Bank for

Reconstruction and Development/The World Bank, 1818 H Street NW, Washington DC 20433. Internet: www.worldbank.org

World Commission on Dams, 2002. Dams and Development - A New Framework for Decision-making. Earthscan Publishing, UK.

Yachiyo, 1995. The National Water Resources Master Plan in The Republic of Zambia. Compiled by Yachiyo Engineering Co. Ltd for the Ministry of Energy and Water Development, Lusaka. Zambia.

ZAMCOM, 2007. The Zambezi Water Course Commission. The Three Year Plan for Operationalisation of the Zambezi Watercourse Commission (2008 - 2011). ZAMCOM Secretariat, Harare, Zimbabwe. Zambezi Watercourse Commission, 128 Samora Machel Avenue, Harare, ZIMBABWE. http://www.zambezicommission.org

Zhou, Y., 2012, Introduction of Modflow - (Lecture Notes), UNESCO-IHE Institute for Water education, Delft, The Netherlands.

ZPWMA (2009) Mana Pools National Park: General Management Plan, Harare, Zimbabwe, p. 1-104.

ZRA, ARA-Zambeze, HCB, ZESCO, 2011. Zambezi River Basin Joint Operational Technical Committee (JOTC) Memorandum Of Understanding. Zambezi River Authority, Lusaka, Zambia. http://www.zaraho.org.zm/

T - #0420 - 101024 - C190 - 244/170/10 - PB - 9781138031807 - Gloss Lamination